NUREG/CP-0190

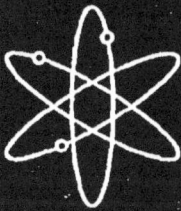

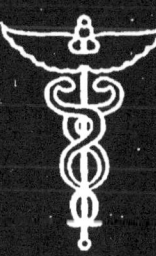

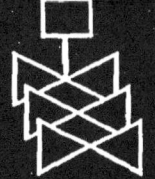

I0482632

Proceedings of the Advisory Committee on Nuclear Waste

Working Group on Performance Confirmation Plans for the Proposed Yucca Mountain High-Level Waste Repository

July 29-30, 2003

U.S. Nuclear Regulatory Commission
Advisory Committee on Nuclear Waste
Washington, DC 20555-0001

AVAILABILITY OF REFERENCE MATERIALS IN NRC PUBLICATIONS

NUREG/CP-0190

oceedings of the
lvisory Committee on
iclear Waste

orking Group on
rformance Confirmation
ins for the Proposed
icca Mountain High-Level
aste Repository

ly 29-30, 2003

iscript Completed: December 2004
Published: December 2004

d by:
 Coleman, Senior Staff Engineer

 Nuclear Regulatory Commission
isory Committee on Nuclear Waste
:hington, DC 20555-0001

ABSTRACT

This report contains the information presented at the meeting of the Working Group on Performance Confirmation Plans for the Proposed Yucca Mountain High-Level Waste Repository. This working group session was convened by the Advisory Committee on Nuclear Waste and held at the Nuclear Regulatory Commission headquarters in Rockville, Maryland, on July 29-30, 2003. This report summarizes the presentations given to the Committee, along with the presentation materials and selected discussions among the participants. The working group included a panel of six experts who observed and commented on the proceedings.

The purposes of the working group were (1) to increase ACNW's technical knowledge of plans to develop and conduct performance confirmation (PC) work for the proposed Yucca Mountain repository, (2) to understand NRC staff expectations for performance confirmation, (3) to review examples of performance confirmation work being planned, (4) to identify aspects of performance confirmation that may warrant further study, and (5) to complement the previous working group session on performance assessment.

CONTENTS

WORKING GROUP ON PERFORMANCE CONFIRMATION PLANS FOR THE PROPOSED YUCCA MOUNTAIN HIGH-LEVEL WASTE REPOSITORY

The U.S. Nuclear Regulatory Commission (NRC) Advisory Committee on Nuclear Waste (ACNW or the Committee) held its 144th meeting on July 29–31, 2003, at Two White Flint North, 11545 Rockville Pike, Rockville, Maryland. The ACNW published a notice of this meeting in the *Federal Register* (68 FR 43238) on July 21, 2003 (Appendix A). This meeting served as a forum for attendees to discuss and take appropriate action on the items listed in the agenda (Appendix B). The entire meeting was open to the public.

ACNW Members who attended this meeting were Chairman Dr. B. John Garrick, Dr. Michael T. Ryan, Mr. Milton Levenson, and Dr. George M. Hornberger. Dr. Ruth Weiner from Sandia National Labs participated as an invited expert.

July 29, 2003

Chairman B. John Garrick convened the meeting at 9:30 a.m. and stated that the meeting was being conducted in conformance with the Federal Advisory Committee Act. He introduced the invited expert, Ruth Weiner and the members of the panel of experts: Chris Whipple, Richard Parizek, John Kessler, Steve Frishman, Robert Bernero, and Wendell Weart. Chairman Garrick then turned the working group over to Vice Chairman Mike Ryan who had the lead responsibility for chairing the working group. Member Ryan reviewed the purposes of the working group and briefly summarized the meeting agenda. He then introduced the keynote speaker, Dr. Chris Whipple (Environ).

Keynote Presentation by Dr. Chris Whipple (Environ)

Dr. Whipple (Environ) gave a keynote talk on "Performance Confirmation for Yucca Mountain." He began by cautioning that the term "confirmation" might not convey the right idea. It may indicate overconfidence, or even suggest that deviations from predictions are failures. Dr. Whipple noted that a good program should be flexible, iterative, risk-informed, and connected to high-level performance goals and should involve the public and increase confidence at each stage. He questioned the extent to which performance confirmation will be needed for conditions and systems that do not bear on compliance. Dr. Whipple noted "traps" to be avoided in a performance confirmation program:

- agreeing to do things that can't be done
- agreeing to measure things that don't affect performance
- claiming safety based on monitoring that is too limited in duration or extent
- requiring unnecessary accuracy or precision in measurements
- failing to establish and apply a system to periodically reconsider performance confirmation requirements

He proposed several criteria for selecting program activities, such as having a threshold of importance that is based on total system performance assessment (TSPA) results and sensitivity studies. It is important to address issues of public concern even if they are not seen as valid by technical experts. However, this should not be an excuse for doing otherwise low-valued work. Dr. Whipple noted that important lessons can be learned from the licensing of the

Waste Isolation Pilot Project (WIPP) at Carlsbad, New Mexico. Performance confirmation should not be used to put off awkward key technical issues. It is important to plan for periodic review of requirements, given that they could change as data become available.

From a practical point of view, Dr. Whipple noted that environmental monitoring of groundwater may provide public confidence but is unlikely to detect early waste package failures. Monitoring of ventilation gases could do so more directly and reliably. Thermal-hydrologic predictions can readily be tested under performance confirmation. However, given that compliance and long-term performance are insensitive to such factors in TSPA models, it isn't clear how this information would be used or whether it could contribute to safety.

Dr. Whipple sees value in continuing corrosion experiments in a way that addresses source term models and parameters. He feels it is unclear whether measurements of the critical metals will confirm or refute the corrosion models, but considers it is important to keep trying. He noted that Joe Payer (Case Western Reserve University) keeps saying that the uncertainty in corrosion is the uncertainty in the environment. It might be possible 5 years into operations to send in a robot to get dust swipes from a waste package canister to determine the chemical starting point for such dust mixtures and whether they differ from the normal desert dust mixed with ground-up Yucca Mountain rock. That might be a creative performance confirmation idea worth doing.

On the subject of contentious scientific issues, Dr. Whipple mentioned the Szymanski hypothesis. In the end, the amount of work that was done helped give people confidence that this differing view wasn't buried by "political muscle." U.S. Department of Energy's (DOE's) recent funding of work at the University of Nevada at Las Vegas on fluid inclusions was a very helpful step in establishing whether Szymanski was right or wrong. In general, it is important to evaluate whether work done to date measures up against the prevailing standards of good science in a particular technical arena. It's not reasonable in any technical arena to say, "Let's wait until 2050 because, undoubtedly, the science will be better then." Has the work been of credible technical content weighed against prevailing standards of good science?

Introduction to Performance Confirmation (NRC's Expectations Regarding Content of PC Plans in a License Application)

Jeff Pohle (NRC staff) gave a talk titled "Performance Confirmation Program, Subpart F of 10 CFR Part 63." This talk was a brief summary of requirements under NRC's site-specific rule. Pohle reviewed general requirements for performance confirmation, including confirmation of geotechnical and design parameters, design testing, monitoring and testing of waste packages, and other requirements. Performance confirmation must start during site characterization and continue until permanent closure.

Introduction to Performance Confirmation

Deborah Barr (DOE staff) gave a talk titled "Overview of Performance Confirmation." She described how performance confirmation focuses on activities designed to confirm the technical basis for the licensing decision. The program should demonstrate that the system and barriers are operating as predicted. DOE has updated its Performance Confirmation Plan for the following reasons:

- to address the requirements in 10 CFR Part 63
- to reflect the barriers that are important to waste isolation

- to use a risk-informed performance-based approach to determine how to confirm each barrier's performance
- to ensure that the program is consistent and compatible with repository operations

In DOE's new vision of performance confirmation, a risk-informed and performance-based approach will be used to determine the complexity, extent, and number of activities to include for testing the effect of a parameter on total system performance or on a particular barrier. The program is designed to be compatible with operations rather than impose substantial design requirements, and it is intended to support an eventual license amendment for repository closure.

DOE has conducted a formal multiattribute utility analysis to determine the relative value of proposed performance confirmation activities. This analysis combined technical judgments with management "value" judgments on the importance of different goals. This multiattribute utility analysis is currently undergoing DOE review and is intended to be released this calendar year in Revision 2 of the Performance Confirmation Plan. Revision 3 of the plan is scheduled for release in the spring of 2004. Revision 3 will:

- define activities
- provide a crosswalk to current and previous testing
- establish the expected baseline for performance confirmation activities
- describe management and administration of the program
- identify needed test plans
- define the process for reporting variances from baseline and describe the appropriate corrective actions

[Note: As of July 2004, neither Revision 2 nor Revision 3 of the Performance Confirmation Plan has been released by DOE].

Decision Analysis Process Used to Develop a Performance Confirmation Program

Karen Jenni (Bechtel SAIC Company; Geomatrix Corporation) described how DOE selected the "portfolio" of tests that will constitute the performance confirmation program. DOE used a formal multiattribute utility analysis to provide a consistent, logical, and defensible basis to compare activities being considered for inclusion in the program. Three criteria were developed to evaluate activities (and measured parameters) being considered for inclusion in a performance confirmation program:

- barrier capability and system performance sensitivity to the parameter
- confidence in the current understanding of the parameter
- accuracy with which the proposed activity measures or estimates the parameter

Technical judgments about sensitivity, confidence, and accuracy were made by the technical experts who were most familiar with the topics. A "core" team of technical experts independently assigned "utility scores" as a consistency check. Where large differences existed in the scores of the technical experts and the "core" team, the scores were discussed and reconciled until differences were small. The few differences that could not be resolved through discussions were reviewed and resolved by a knowledgeable senior manager. Costs of various activities were also considered in developing test portfolios. DOE initially considered 237 parameters and 360 activities for possible inclusion in portfolios. Altogether, DOE developed 11 portfolios, of which 6 were evaluated in detail. The portfolio designated C has been selected by DOE's BSC (Bechtel SAIC Company) Manager of Projects and senior advisors as a starting point for the performance confirmation program. This program was considered to be cost-effective and captures 82 percent of the total potential utility. Portfolio C underwent further review by BSC senior management. Of the original 99 activities, 26 were removed because they were more logical candidates for other testing programs, 3 activities were combined with other activities in the program, 3 were retained in principle but modified in scope, and 2 new activities were added. BSC then proposed the resulting modified portfolio to DOE.

Elements of a Performance Confirmation Program—A Presentation of DOE's Selected Program and Its Components

James Blink (Bechtel SAIC Company; Lawrence Livermore National Laboratory) gave a talk titled "Elements of the Yucca Mountain Performance Confirmation Program." He cautioned that some changes may occur during DOE's acceptance review and as the activities are developed for a license application. Dr. Blink noted that Phase 1 of the decision analysis was risk-based because it relied on performance assessment calculations. It was performance-based because it considered performance of the individual barriers and the total system. Phases 2 and 3 are considered risk-informed because they consider other factors such as relationships among activities, feasibility, operability, and cost.

The decision analysis focused the performance confirmation activities on the areas of highest risk. Three main groups (or classes) were identified:

- Disruptive scenario classes—igneous activity and seismic activity scenarios
- Biosphere-related activities—applicable to multiple scenarios
- Nominal scenario class (lower risk than the disruptive scenarios)—waste package and drip shield, pre-emplacement environment, land surface characteristics and the unsaturated zone below and above the proposed repository, coupled thermal processes, saturated zone, and cladding/waste form/invert

Igneous activity is the largest single contributor to the probability-weighted annual dose to the reasonably maximally exposed individual. The approximately 13 related performance confirmation activities will be designed to confirm the assumptions, data, and analyses of igneous events. Specific work may include:

- drilling of aeromagnetic anomalies to investigate possible buried volcanoes
- an updated expert elicitation on the probability of an igneous event
- further analyses of igneous consequences
- monitoring of regional extensional tectonics

Twenty-two activities have been proposed for performance confirmation of the waste package and drip shield. These will investigate mechanistic details of waste package and drip shield corrosion, and will perform lab tests on mockups to confirm stress sources as consequences of rockfalls and seismic activity. The near-field environments will be studied in thermally accelerated drifts using drift-end instruments, in-drift sampling, and a remotely operated vehicle. Parameters to measure will likely include temperature, humidity, dust and gas composition, pressure, radiolysis effects, condensate chemistry, thin film chemistry, and microbial activity. Radionuclides will be monitored in exhaust air to detect any waste package breach. The pressure seal of all waste packages can be measured with the remotely operated vehicle. This vehicle can also inspect emplacement drifts for ground support integrity and "shape."

Other performance confirmation activities will relate to seismic activity (3), the biosphere (6), the pre-emplacement environment (8), coupled thermal processes (~12), and elements of unsaturated and saturated hydrology and chemistry (~13).

Documentation and Further Development of the Performance Confirmation Program—A Presentation on Possible Changes in the Next Revision of DOE's Performance Confirmation Plan

Deborah Barr (DOE) described how the PC program will likely evolve. Her talk was titled "Documentation and Further Development of the Performance Confirmation Program." Revision 3 of the Performance Confirmation Plan is scheduled for spring of 2004. This revision will include specific details about the plans, including:

- specific activities (what, when, where, and how)
- baseline established for PC
- bounds and tolerances for parameters
- management and administration of the PC program
- test plans
- the process for reporting variances and appropriate corrective actions

Some of the proposed PC activities will require feasibility evaluation or even the development of new technology.

Public Comments (Day 1)

Judy Treichel (Nevada Nuclear Waste Task Force) noted a problem with some of the meeting handouts. She was also concerned that there appears to be no performance confirmation program. It was supposed to have started during site characterization. She also wondered if Rev. 2 of the plan is being planned, what happened to Rev. 0 and Rev. 1. Ms. Treichel was concerned about how an expensive performance confirmation program would be paid for. Have there been any demonstrations of retreivability equipment such as remotely operated vehicles? Would they be able to work with all the heat and radioactivity present in the future underground. Some of the key technical issues now appear to be referred to as less important. But at one time they had to be resolved.

July 30, 2003

NRC's Risk Insights Initiative and Its Impact on Review of Performance Confirmation Plans

Timothy McCartin (NRC) gave a talk titled "Risk-Informing Performance Confirmation." DOE is required to identify and describe repository barriers. Under a risk-informed approach, DOE would identify the relative risk significance of each barrier. Mr. McCartin described the approach as an iterative one in which risk significance is described, a quantitative basis is provided, uncertainties are considered, important parameters and assumptions are identified, and confirmatory evidence is considered.

Mr. McCartin gave a hypothetical example indicating that retardation in alluvial deposits is risk significant. Alluvium has the potential to delay the movement of most radionuclides for very long time periods. There is little uncertainty about the retardation of I-129 or Tc-99. These radionuclides are highly mobile and move with the water. Am-241 and Pu-240 tend to be relatively immobile under most circumstances. However, the retardation factor for Np-237 is highly variable such that this radionuclide could be retarded for a few centuries or more than 100,000 years.

Mr. McCartin noted that risk insights identify areas to be considered for performance confirmation. An NRC risk insights report is now being prepared that is based on a risk baseline, provides quantitative bases for relative risk, and identifies further calculations that may be needed. The risk insights report will be updated in the future as needed.

NRC's Acceptance Criteria in the Yucca Mountain Review Plan (YMRP) for Review of Performance Confirmation

Jeff Pohle (NRC) gave a talk titled "Performance Confirmation Program - Section 2.4 of the Yucca Mountain Review Plan." He noted that the review plan has four primary review areas. The first area consists of the general requirements, including the objectives to acquire data to indicate whether subsurface conditions are within assumed limits and whether barriers are functioning as anticipated. This first area also includes schedules, provision of baseline information, and the monitoring and analysis of possible deviations from baseline.

The second area of review deals with the confirmation of geotechnical and design parameters, such as the insitu monitoring of the thermomechanical response of the underground facility until permanent closure. The third review area addresses design testing, including, for example, tests of borehole or shaft seals and drip shields and the evaluation of thermal interactions of engineered barriers with the natural environment. The fourth review area concerns the monitoring and testing of waste packages.

Mr. Pohle noted that to achieve an adequate review of performance confirmation, NRC reviewers will need to be familiar with:

- barriers important to waste isolation and any unresolved NRC concerns
- DOE's description of the capability of each barrier to isolate waste
- DOE's information on uncertainties related to parameters, processes, models, etc., for each barrier
- DOE risk evaluations and NRC's risk insights baseline
- CNWRA support to enhance independent review capability.

Mr. Pohle stated that NRC needs an educated staff that is knowledgeable about DOE's description of what the barriers are, what the capabilities for the barriers are, the outstanding concerns in these areas, information about uncertainties, the evidence related to these parameters, and information from NRC-generated risk evaluations. Support from NRC's technical assistance contractor, the Center for Nuclear Waste Regulatory Analyses (CNWRA) will be needed to help enhance NRC's capability to independently review performance confirmation. CNWRA is currently doing work in the area of instrumentation, looking ahead at the types of testing activities DOE may propose to do and the instrumentation required. CNWRA is also looking at longer-term tasks on software requirements for future changes in computer codes, particularly thermohydrologic codes.

John Kessler, Electric Power Research Institute (EPRI), observed that there seems to be a disconnect between what NRC is emphasizing in performance confirmation and "almost everything else." He heard from NRC speakers an emphasis on every barrier, regardless of its individual contribution to overall performance. If DOE calls something a barrier, it appears NRC is going to ask them to defend it equally, whether it is the waste package or whether it is the saturated zone. Dr. Kessler noted that DOE considers the saturated zone to be relatively unimportant, but NRC considers it to be important. It appears that the two organizations are taking fundamentally different approaches, and this relates not only to performance confirmation but to the whole license application.

Mr. McCartin responded that NRC is looking at the potential to contribute to overall risk. For example, neptunium tends to be the largest dose contributor. In the natural system, the alluvium has the potential to significantly retard the most important radionuclide for overall risk. And that's why, with regard to neptunium, the saturated zone (specifically alluvium) is important.

Presentations by Representatives of the State of Nevada, Several Affected Counties, the Las Vegas Paiutes, and the Electric Power Research Institute

Les Bradshaw (Nye County, Nevada) gave a talk titled "Nye County's Views on Performance Confirmation and Related Topics." He noted that PC is a critical program element because it will show whether the repository will perform in a way that protects health and safety in Nye County. Mr. Bradshaw was concerned that no approved program appeared to be in place. He expressed concern that DOE suspended monitoring of several unsaturated zone boreholes in 2001. He felt that this monitoring should have been a part of a PC program.

Mr. Bradshaw considers that a comprehensive PC program should have been in place long ago and that Nye County, Nevada, and other stakeholders should have had a chance to review it. PC should include significant participation by qualified groups from outside of DOE. PC will be more acceptable to the public if some of the work is done by qualified independent groups. The following PC tasks could be undertaken by independent entities: (1) technical review of plans, data, and analyses; (2) establishment of baseline data for water, air, rock and soil, and biota; (3) post-emplacement monitoring of the environment; and (4) storage and dissemination of PC data. Nye County is already participating in PC work that will be related to Nye County's Early Warning Drilling Program. DOE has approved funding for expansion of this work through 2007.

John Walton (University of Texas at El Paso, consultant to Nye County) gave a talk titled "Some observations on performance confirmation and performance assessment." Nye County has several areas of concern, including the anticipated impacts of a repository on Nye County resources and potential unresolved performance assessment issues. Dr. Walton suggested that the future heating up of the mountain will cause the top of the mountain to become warmer and wetter, resulting in possible changes in flora and fauna. These changes could take place in tens to hundreds of years. Therefore, soil conditions and vegetation changes should be monitored over time. A baseline of vegetation communities should be obtained before a repository is built.

Dr. Walton observed that tunnel roof collapse remains an unresolved question, because rubble would act as insulation and change conditions assumed in coupled thermo-hydrologic modeling. Backfill may be needed to provide a predictable environment. Dr. Walton was concerned that the natural ventilation of the mountain may not be fully accounted for in DOE models. This is important for heat, moisture, and chemistry modeling. He also stated that DOE's models mix spatial and temporal variability with uncertainty, which can unrealistically spread projected risk and reduce peaks in mean projected doses. Dr. Walton wonders if this mixing of variability and uncertainty is conservative or nonconservative in the context of Yucca Mountain.

Dr. Walton gave an example of a simplified "pseudo" performance assessment that included four processes: corrosion, release rate, transport lag time, and an unspecified event that fails the remaining waste containers when it occurs. He compared two simulations. In one simulation he took the mean dose representing 1,000 Monte Carlo realizations. The results are compared to a second simulation, where the standard deviation is increased for one parameter, which increases the uncertainty range. Contrary to expectation, in this latter case the risk is actually reduced because it is measured as the peak of the mean of the realizations. What happens is, sometimes when you modify a parameter, each of the 1,000 realizations will have its peak occur at different points in time. That is, the peaks of the individual realizations will be spread in time. So when the mean is calculated, it broadens and flattens relative to the curve with less variance. The projected risk is lower, and performance has actually been improved by ignorance. This is not a general conclusion, because if different parameters are changed, sometimes the risk increases. The results depend on which parameter is broadened. It's complicated and not obvious what the result will be. Therefore, in performance assessments for Yucca Mountain, when are so-called "one-off" and "one-on" analyses conservative or nonconservative? Dr. Walton described a scenario in which a DOE manager is asked to fund a study on the sorption coefficient (K_d) of neptunium. Will the manager really want to fund it if credit is being taken for the fact that the K_d isn't well known? In conclusion, Dr. Walton noted that local involvement is crucial to performance confirmation because otherwise the work is the product of an internal "group think" and doesn't produce as many ideas. Dr. Walton stated that Nye County should be involved.

Mr. Steve Frishman (State of Nevada) commented that the performance confirmation requirement and its meaning are essentially identical to what was in Part 60. He was concerned that it looks like performance confirmation has been analyzed out of the regulation by the Yucca Mountain Review Plan. He reviewed the definition of performance confirmation. It is a program to confirm the validity of the information that is used to support the reasonable expectation determination. It's to begin during site characterization and continue through closure. Mr. Frishman said that if you put it in the context of the regulatory process, it seems like its purpose is a simple one. That is to provide some additional confidence in the technical basis for a decision to amend the license for closure. Under the regulation, the disposal decision is made with the construction authorization decision. And everything after that would be amendments in one way or another, but they need to be supportive of that original disposal decision.

Mr. Frishman sees a "danger" of unfinished business in site characterization being casually flipped into the performance confirmation "bucket." He sees the project in a situation where there are areas where site characterization is not complete. But, at the same time, there is the recognition that the license application has to be one that is adequate for a decision regarding reasonable expectation that the performance requirement will be met. Mr. Frishman would be greatly concerned if there were any approach literally on the part of anyone to use performance confirmation to overcome this incomplete site characterization and actually get to a point where it gains significance in licensing. He believes that the license application review and the hearing should proceed to a reasonable expectation decision without any deference whatsoever to the substantive content of the performance confirmation program. Performance confirmation is essentially an add-on. And it should have literally no basis in the disposal decision that comes at the time of a decision on construction authorization. PC is a good thing to do. But it should be given no deference in licensing. What Mr. Frishman sees coming is making a lot of things into license conditions hooked into this vehicle of performance confirmation such that in effect site characterization never ends.

Performance confirmation should help take a hard look at the performance approach that has been taken and maybe not think so much in terms of looking at what is most important, not sort of doing endless reiterations and rethinking about the components of the waste package model. The most important thing is to go back and challenge the conceptual models on which the performance assessment is built.

Mr. Frishman cautioned that this workshop should not be used as a means to try to revisit what performance confirmation could be if it were to be most friendly to a license application, most friendly to the applicant, or maybe even most utilitarian to the regulator. Performance confirmation is a pretty simple thing to use in a common sense way, not in a way that results in an uncertain job only becoming more uncertain because someone found it to be convenient because it is the only bucket left out there to throw stuff into.

Dr. Atef Elzeftawy (consultant) presented remarks to the Committee on behalf of the Las Vegas Paiute Tribe. He discussed a number of topics, including the need for regulators to be tenacious in finding out how DOE plans to do the work. He had discovered years earlier that DOE was planning to drill unsaturated zone holes using drilling mud. That wouldn't have worked and DOE changed its methods. He emphasized the importance of unsaturated zone hydraulic parameters. Dr. Elzeftawy also suggested that the tribe had a concern that funding for things like PC might initially grow, but later dwindle to almost nothing. He submitted to the Committee an article about the golf resort owned by the tribe. This article has been placed in the formal record of the meeting.

Mr. Engelbrecht von Tiesenhausen (representing Clark County, Nevada) spoke on the topic "Performance Confirmation, what does it really mean?" He discussed the general requirements and definitions of PC that are described in NRC's Part 63 regulation. Mr. von Tiesenhausen noted that there are several challenges to PC, such as estimating temperature effects in a repository, and the idea that even in a tunnel dedicated for PC, conditions are unlikely to reproduce those found in a repository. Clark County considers waste package performance to be the most critical performance issue. Long-term corrosion data in a representative environment is "most likely impossible" to collect before a repository would be closed. Mr. von Tiesenhausen stated that PC should not be used to put off the resolution of issues that are part of a license application. PC should confirm results but not be a primary source of data. Any license application that relies on PC and formal requests for additional information should be looked at very critically. Mr. von Tiesenhausen suggested that PC studies can help us better understand the natural system in several ways. For example, such work can improve the understanding of the role of the Calico Hills geologic formation on waste isolation. It can help to better interpret where and how fast water travels in the natural system. And finally, PC studies can improve the understanding of current and future geochemical processes.

Dr. Kessler gave a talk titled "The role of performance confirmation in Yucca Mountain development." He described differences between performance confirmation and long-term research and development. PC is specifically designed to evaluate the technical bases for the licensing decision. EPRI has performed several activities related to PC, including evaluation of DOE's draft 2000 report, convening of a PC panel to make recommendations, hosting of a PC workshop in 2001, and documentation of the above in a December 2001 EPRI report. Dr. Kessler recommended that NRC and DOE start now to develop a shared understanding of how both PC and long-term monitoring will be carried out. A flexible plan is needed, with work activities to be prioritized using risk-informed judgment. He noted that NRC and DOE have made a commendable start, NRC with its risk-informed regulation and DOE with an initial PC Plan.

Dr. Kessler described possible criteria for prioritizing activities, such as (1) risk-informed, (2) timing of the need for data, (3) cost of an activity, (4) interference with other activities, (5) stakeholder agreements and stakeholder concerns, (6) health effects to workers and the public, and (7) ability to define the activity in such a way that "confidence" would be enhanced. Traps to be avoided include agreeing to measure things that do not affect performance, agreeing to do things that can't be done, requiring unnecessary accuracy or precision, monitoring for too short a time, assigning excessive levels of conservatism, and neglecting the need to maintain technical capabilities.

Dr. Kessler suggested a number of options to address important FEPs (features, events, and processes) that are not amenable to PC testing. These options are to (1) use reasonably bounding values based on expert elicitations, (2) leave a reasonable margin, (3) use natural analogues, and (4) add or modify engineered features to reduce the importance of the FEPs. These types of FEPs should be identified early.

Dr. Kessler advised that meaningful tolerance bands need to be established now, that a clear beginning and end must be defined for PC activities, that appropriate "baseline" information must be collected at the right times, and finally, that activities should be prioritized in case of limited funding or time.

Research Perspective on Long-Term Testing for Performance Confirmation— Development of an Integrated Groundwater Monitoring Strategy

Dr. Thomas Nicholson (NRC Office of Nuclear Regulatory Research) reviewed the ongoing development of an integrated groundwater monitoring strategy from a generic research perspective. The objectives of this research are (1) to develop technical bases for NRC staff evaluation of groundwater monitoring programs, (2) to couple monitoring to site characterization and facility performance assessment, and (3) to assess monitoring strategies to identify and support relevant alternative conceptual models of flow and transport. Other research objectives were (4) to identify hydrologic performance indicators, (5) to develop a design strategy to collect monitoring data for parameter estimation, model calibration, and uncertainty analyses, and (6) to transfer technology to the NMSS staff.

Working Group Roundtable Panel Discussion on Performance Confirmation

Dr. Whipple noted that Part 63 requires PC for all barriers that are classified as important to safety and that the PC work must be practicable. He suggests there is potential conflict between the two requirements, and he believes there is a possibility that DOE has not prioritized well and has failed to see the downside to classifying so many things as important to safety. Mr. McCartin (NRC) responded that DOE has some flexibility in deciding which barriers it will rely on. There is no numerical value given to describe the significance of barriers, but NRC would expect the DOE to evaluate the most significant barriers in developing their safety case. In looking at PC, DOE would also be looking at the barriers it is relying on most. Dr. Whipple wondered what NRC would do if DOE identified a larger number of barriers than a reasonable person might technically believe are important. Would NRC rescue DOE from its own folly? Mr. McCartin replied that NRC is not there to "rescue" DOE. He referred to NRC's review plan for post-closure performance and noted that it emphasizes up front the identification of barriers important to performance. The intent is to tailor the NRC review to closely examine the barriers that DOE relies on the most. Generally, an NRC review focuses on what hasn't been considered or has been left out.

Robert Bernero (NRC, retired) observed that this is a classic problem in nuclear licensing involving the NRC. The applicants for a license are chronically looking for a prescriptive formula, "Tell me what I need to do so I can do it and you'll therefore give me a license." And the staff is chronically trying to give a description, an approach, but the responsibility for the logic and the supporting programs is the applicant's. That's a common problem, and especially so for DOE because the DOE is not accustomed to conducting its affairs as a regulated licensee.

John Garrick (ACNW Chairman) stated that the issue of classifying something as safety- or non-safety-related is reminiscent of an analog used in probabilistic risk assessments, i.e., the "rocks in the pond" example. You have a pond that has a lot of rocks sticking out, and when you remove the biggest rock, the pond level goes down a level and some more rocks surface, and finally you remove enough rocks and the remaining rocks are small enough now that the surface doesn't significantly change as they are removed. That's what the performance assessment is supposed to give you. The answer to the question of whether or not it's safety important is whether it makes any difference to the bottom line. If the performance assessment was competently prepared, there will be a road map that says "I'm not going to measure or worry about this particular rock because no matter what I do with it doesn't change the performance, it doesn't change the lake level of the "pond." If we have any confidence in our analysis at all, we have an inherent mechanism for classifying whether it's safety important or not, whether we need a particular barrier or not, whether it contributes to performance or not.

Steve Frishman (representing the State of Nevada) discussed, as an example, the parameter of matrix diffusion. Years ago the DOE had decided not to take credit for it because it was worth a relatively small percent of performance. It is also relatively unimportant in NRC's model. DOE seems to be reconsidering the potential contributions of such parameters. Mr. Frishman supported the idea that if a parameter is not worth a lot to performance for an applicant, to avoid an onerous review process, don't take credit for the parameter in the first place.

Richard Parizek (Penn State and Nuclear Waste Technical Review Board) stated that he was speaking as a private citizen rather than as a board member. He mentioned that some very valuable lessons were learned at the WIPP, that there is a real program there. There's an opportunity to understand how that program worked and why those decisions were made to include or not include certain testing efforts. There's a lot to be said about what we need to know about a site and about the characteristics of the site. For instance, what is assumed about climate in the TSPA model? Look at the Death Valley area (California) and the Mojave River drainage basin and we see in 10,000 years four major lake level stands in the basins. There were several periods of alluvial fan development, which really requires big triggering mechanisms to flush sediments down to generate the fans. So there's something about this weather story and about monitoring that might then say, "I'd better start looking underground because maybe this is a time when fast paths will kick in and this may have something to do with repository behavior." Mr. Parizek noted that from a science understanding point of view and confidence building point of view, some people wouldn't care where the money came from as long as PC got done. He discussed a number of possible monitoring activities, such as the placement of a monitoring well to monitor water chemistry and groundwater elevations and the drilling of magnetic anomalies to try to detect buried basalt flows.

Mr. Bernero asked, "What shall the program pursue in performance confirmation testing?" He noted that barriers should be tested, but unimportant barriers may not be. They may be set aside, but important performance assessment models may call for resurrecting. The key thing is to test models and the performance assessment. The performance confirmation program, the entire safety analysis, has to be a living system, a living document, learning and incorporating that learning and changing accordingly. It is important in any program to look at those things that provide important support for performance assessments, but that's not quite all you want to do. What is needed is to go beyond trying to measure things that can confirm that performance, and look broadly enough to find any holes or differences in models or assumptions that may surround those models and techniques that you believe to be correct. Usually surprises come in finding things that we didn't expect, and PC as a tool ought to be broad enough to look for those kinds of things.

Wendell Weart (Sandia National Labs, Senior Fellow) spoke about his WIPP experiences, and noted that DOE sometimes promised to do things that they weren't able to do. He expressed hope that PC wouldn't become a "shopping basket," that work activities would be selected carefully based on what is really important. Mr. Weart noted that this is a program that's going to be long enough that early on there may be intense interest and funding for it, but in future funding may lag, making it a continuous struggle to implement the program. Regarding the use of conservative bounding arguments, Mr. Weart found from his WIPP experience that programs of long duration can be hurt by the fact that bounding conservatisms have been locked in, and are very hard to change after the fact. He advised not adopting these conservatisms unless it really is necessary.

Dr. Whipple also commented on the idea of avoiding bounding analyses and trying to be as realistic as one can be. Regulators find enormous comfort in being handed a bounding analysis with a lot of margin. There's little chance of that coming around and biting them. Dr. Whipple thinks this could similarly be true for a 9-million-page license application to the NRC. He noted that one aspect of a fully realistic analysis is it represents best understanding, best estimates with a kind of a 50-50 chance of being wrong in the nonconservative direction. This may be unacceptable in a politically charged, politically visible licensing process. As desirable as it would be to have a fully risk-informed approach through the licensing process, it would be a very risky strategy for an applicant to take. There is intellectual merit in a risk-informed approach, but the political reality of a licensing approach is the burden is on the applicant to prove that everything they say is either true or wrong in the safe direction. That is not fully compatible with being realistic and risk-informed.

Mr. Bernero responded that NRC, in its approach to a probabilistic risk analysis for reactor plants, made a concerted effort to be realistic, but approached realism from the conservative side of the field. There was simplification. For example, if conditions for adequate core cooling are lost, it was assumed that the core melted right away rather than try to mechanistically model the whole process. There was a very important reason why that could be done in a regulatory environment. Mr. Bernero noted that NRC consciously avoided regulating with a safety goal. It described a safety goal, one-tenth of 1 percent increment of background risk, etc., but did not regulate to the safety goal. It was intended to use performance assessments, or probabilistic risk assessments (PRAs), that were as realistic as they could be made. The big difference regarding high-level waste is the fundamental basis of the regulation is to regulate with the performance assessment. It's not a safety goal, it's a condition of acceptability. The real question is trying to understand the margin, trying to understand what confidence you can have in those results, and trying to understand barriers that right now may not be very important, but if the principal barrier of the package, etc., fails, they become very important. Mr. Bernero considers there's a fundamental difference in NRC history in that regard.

Mr. Frishman responded that some people have suggested that performance assessment should be an exposure of what you know. It should be possible to accurately characterize and quantify what you don't know. On the other hand, a performance assessment has to be used for compliance because that's what the rule says. Mr. Frishman suggested there may be the need to develop an expectation for two kinds of performance assessments. One of them will meet the need required by the rule to demonstrate what you know, and the other will show compliance based on an assessment of a demonstration of what is known.

Dr. Weart commented that one can't always judge in which direction conservatism exists. And if you're smart enough to have thought of everything in advance and say, "I'm never going to have any surprises," then perhaps you're okay. Dr. Weart advised that if you don't have to rely on bounding analyses, don't, but there are times when perhaps it's all right. But it can come back to haunt you.

Mr. Bernero commented on DOE's decision analysis for selecting the PC portfolio. He found the decision analysis process difficult to track but clear, and thought it was very well done, a logical process, clearly tracked, and producing a reasonable result. However, he found some of the characterization of portfolios A through K to be unclear. Portfolio A was identified as the minimum needed to satisfy the regulator. Mr. Bernero felt that wouldn't be right because that would be the minimum necessary. The applicant would be saying, "I know all I have to do is tell them this, and that's enough to satisfy them." He interpreted DOE's selected portfolio C "plus" as representing the best judgment of the applicant. It is DOE's responsibility to come up with the right performance confirmation, to show how they're going to satisfy the regulatory requirement. NRC would review that, and that sounds like the right way to choose a portfolio. Mr. Bernero commented that the NRC avoids, and should avoid, overly prescriptive regulation. NRC shouldn't give DOE a prescriptive description of what the performance assessment should be. But NRC should develop alternative models of their own. They should be giving descriptive analyses that say what the performance confirmation ought to be.

Dr. Garrick stated that the regulator is never the expert on the system being licensed that the operator-owner is. Never. No matter how many regulations, no matter how many lawyers the regulator has, the regulators do not know the system as well as the owner, operator, designer, builder, or whoever. The perspective should be that the most expert group in the world on that system is completely satisfied that the system is safe. They shouldn't even think compliance—they should think totally from the standpoint that it's safe, and then let the licensing people worry about whether they've complied with the regulations. Mr. Bernero agreed that the regulators are not the ultimate experts, and regulations cannot be so prescriptive as to have specific solutions to problems. But they can require a competent quality assurance (QA) program. He remembered signing a letter on July 31, 1989, to the Yucca Mountain program that said, "This won't wash. Your site characterization plan is—we have two objections to it. You don't have an adequate QA program, and you don't have an adequate design control process." NRC did not tell them what those processes had to be. But DOE was told that what they had didn't "cut the mustard." The regulator can't pose as the expert, but the regulator can say, "You don't meet the standards or evidence. You don't show evidence of sufficient safety or competence in an area."

Dr. Kessler commented that, since Yucca Mountain is a first of a kind project, it's probably okay for there to be a bit more guidance from NRC, given that this is the first one out of the starting block. This doesn't mean a lot more specification, but perhaps some clarification of the relative importance of supporting the barrier analysis versus just supporting the overall performance criteria. Perhaps DOE needs to back up and add a little bit more on the realistic side to provide some insight on how much margin they're providing in their compliance-based assessment.

A number of participants discussed the manner in which NRC would review DOE's PC Plan, which would include discussions in public meetings about what is reasonable for the program to include. Dr. Kessler commented that this dialogue needs to begin now. Dr. Parizek commented that "it's not collusion, it's trying to be efficient with the use of everybody's time and getting to the end point. Mr. Frishman expressed the concern that it will be a very difficult situation if the applicant and the regulator are essentially negotiating the meaning of the regulation. He suggested that there is no real precedent for this. Mr. Frishman felt that to do the informal negotiation prior to licensing could be antithetical to an accountable regulatory system. Dr. Kessler responded that there seems to be plenty of precedent for the regulator and the applicant to have discussions on a generic basis. He gave examples of very quantitative, specific interim staff guidance that grew out of technical discussions in publicly noticed meetings where the applicants and the regulator sat down and talked about a technical detail. Dr. Kessler considers that this happens all the time, and it's done in public meetings with that kind of level of discussion.

Public Comments (Day 2)

Ms. Judy Treichel (Nevada Nuclear Waste Task Force) was very concerned that the PC program is not far better defined at this time. This is one reason that the site recommendation and sufficiency letter were premature. Yucca Mountain is a project forced on an unwilling host. These are people (Nevadans) who do not like the idea of being the host for the repository, and they really don't like DOE. These nuclear-testing people killed us once; we're silly if we let them do it again. And Nevadans have been told for years and years, you don't have to like DOE, you don't have to trust DOE, because you've got NRC. NRC is going to come in here—they will only license this thing if it's absolutely safe, and NRC will take charge of your safety, your health, and your well-being. So be clear about that. That's what has been told to Nevadans, and that's what their expectations are. And you've got people who are very nervous. We don't want to see compromises. You already know the lay of the land in Nevada. But don't let this thing become some sort of an excuse. Ms. Treichel is eager to see what performance confirmation winds up being, but doesn't want it to be something that just hangs over everybody's head.

Dr. Elzeftawy said that the performance confirmation program needs to be simple but beautiful for the people to have confidence that this program is on track and is applicable. He noted that we, as scientists, discuss these issues but the public has some common sense and needs to understand the simplicity of performance confirmation. The NRC has the responsibility of looking at it. But NRC needs to come to a focal point, and the focal point is to make it simple and understandable to most people. Dr. Elzeftawy wondered what DOE has to show for the large expenditure of funds to date. He advised the ACNW to hold more meetings in Las Vegas because of the expense of travel for most citizens.

ABBREVIATIONS

ACNW	NRC's Advisory Committee on Nuclear Waste
ACRS	NRC's Advisory Committee on Reactor Safeguards
CFR	Code of Federal Regulations
DOE	U.S. Department of Energy
DWM	NRC's Division of Waste Management
EPA	U.S. Environmental Protection Agency
EPRI	Electric Power Research Institute
NMSS	NRC's Office of Nuclear Materials Safety and Safeguards
NRC	U.S. Nuclear Regulatory Commission
PA	Performance Assessment
PC	Performance Confirmation
SAIC	Science Applications International Corporation
SNL	Sandia National Laboratories
TPA	NRC's Total-System Performance Assessment
TSPA	DOE's Total System Performance Assessment

1. INTRODUCTION

An Advisory Committee on Nuclear Waste (ACNW) working group on Performance Confirmation of the potential Yucca Mountain High-Level Waste Repository was convened during July 29-30, 2003. This was a technical session on plans to develop and conduct performance confirmation work. The NRC regulations governing performance confirmation for Yucca Mountain are defined in 10 CFR 63, Subpart F. Relevant sections are shown below:

63.131 sets forth the general requirements that subsurface conditions be within the limits assumed in the licensing review and that natural and engineered systems function as intended and anticipated. The program must include monitoring, laboratory and field tests, and in situ experiments as appropriate.

63.132 sets forth the requirements for confirmation of geotechnical and design parameters. This section specifies that, during repository construction and operation, information will be gathered to confirm that geotechnical and design parameters are maintained, that NRC will be informed of any needed changes in them, and that the in situ thermo-mechanical response of the facility must be monitored until closure to ensure the engineering and geologic features are within design limits.

63.133 sets forth requirements for testing during the early stages of construction. A program is required that will test engineered systems and components used in the repository design, e.g., seals, backfill, drip shields, and testing thermal effects of waste packages and backfill (if any) on rock and on groundwater in the unsaturated and saturated zones.

63.134 sets forth the requirements for monitoring and testing of waste packages. This program must be established for monitoring the condition of waste packages in an environment representative of the environment in which the wastes are to be emplaced. Further, the program must include realistic laboratory experiments that focus on the *internal* conditions of the waste packages. The waste package monitoring program must continue as long as practical *up to the time of permanent closure.*

The working group was formed to review the process for planning performance confirmation work, and to review specific examples of studies that could comprise the DOE performance confirmation program. Consideration of key parameters and assumptions (i.e., "pinch points") presented in the TSPA and incorporated within a license application can assist development of the performance confirmation program. This working group was complementary to the earlier working group on performance assessment. Like the Performance Assessment Working Group, the Performance Confirmation Working Group emphasized those activities that are planned to increase confidence by confirming assumptions and conditions used to estimate repository behavior. Of particular importance are assumptions to be made about the mobilization of the radioactive materials in the inventory that are the most important drivers in the performance of the proposed repository.

Purposes

The purposes of the working group were (1) to increase ACNW's technical knowledge of plans to develop and conduct performance confirmation work for the proposed Yucca Mountain repository, (2) to understand NRC staff expectations for performance confirmation, (3) to review examples of performance confirmation work being planned, (4) to identify aspects of performance confirmation that may warrant further study, and (5) to complement the previous working group session on performance assessment. The results of this working group were used to develop a letter report to provide observations and recommendations to the NRC about performance confirmation.

Stakeholders and members of the public were given opportunities to provide input to the proceedings. The working group promoted discussions on:

1. DOE's use of risk insights to develop performance confirmation plans

5. Issues identified by the TSPA/TPA working group, which should be included in a performance confirmation program

6. The activities (i.e., measurements, analyses, and interpretations) of performance confirmation

7. Approaches for analyzing and interpreting performance trends, how action levels or trigger points could be developed, and how DOE would potentially respond to an adverse trend

8. NRC's acceptance criteria for review of performance confirmation, and generally how they will be applied

General Approach

The format of the working group included (1) a keynote presentation by a distinguished scientist or engineer not directly connected to the NRC and DOE performance confirmation work to set the tone for discussions on performance confirmation, (2) a series of expert talks from senior planners of the performance confirmation work itself, (3) talks by NRC staff about regulatory requirements and application of risk insights and acceptance criteria, (4) talks by representatives from units of affected local government and stakeholders presenting their views regarding performance confirmation, (5) a panel discussion of issues and results presented, and (6) public comments.

Representatives from several Nevada counties, the State of Nevada, EPRI, and a Native American tribe presented their views on the general approach and plans for developing performance assessment work. Following the presentations, a panel of experts moderated by the chairman of the working group reviewed the material presented and offered their observations and recommendations on the performance confirmation planning work done to date. Several opportunities were given to participants, including those in the audience, to make comments consistent with the purpose and objectives of the working group session.

We thank all the speakers and participants for their contributions, which collectively made this working group a success. We thank Neil Coleman, Mike Lee, Howard Larson, and the staff of the ACRS/ACNW Operations Support Branch for their efforts in support of the working group.

B. John Garrick, Chairman,
ACNW

Michael T. Ryan, Vice Chairman,
ACNW

George M. Hornberger, Member,
ACNW

Milton Levenson, Member,
ACNW

Left blank intentionally.

4

2. LETTER TO

THE HONORABLE NILS J. DIAZ, CHAIRMAN
U. S. NUCLEAR REGULATORY COMMISSION

FROM

B. JOHN GARRICK, CHAIRMAN
ADVISORY COMMITTEE ON NUCLEAR WASTE

COMMENTS ON WORKING GROUP SESSION ON PERFORMANCE CONFIRMATION FOR YUCCA MOUNTAIN
JULY 30-31, 2003

Left blank intentionally.

.

October 1, 2003

The Honorable Nils J. Diaz
Chairman
U. S. Nuclear Regulatory Commission
Washington, D.C. 20555-0001

SUBJECT: WORKING GROUP SESSION ON PERFORMANCE CONFIRMATION FOR
           YUCCA MOUNTAIN

Dear Chairman Diaz:

During its 144th meeting on July 29-30, 2003, the Advisory Committee on Nuclear Waste
(ACNW or the Committee) held a working group session (WGS) on performance confirmation
(PC) for the proposed high-level waste repository at Yucca Mountain, Nevada. PC refers to the
tests, experiments, and analyses that will be performed to evaluate the adequacy of the
information used to show compliance with performance objectives in 10 CFR Part 63.

The purposes of the WGS were to (1) increase ACNW's technical knowledge of plans to
develop and conduct PC work, (2) understand NRC staff expectations for PC, (3) review
examples of PC work being planned, (4) identify aspects of PC that may warrant further study,
and (5) complement the previous working group session on performance assessment. The
WGS included a panel of six distinguished experts from academia and various government and
private institutions. Representatives of the U.S. Department of Energy (DOE), the U.S. Nuclear
Regulatory Commission (NRC), and the State of Nevada made presentations, as did various
other stakeholders.

DOE's PC program is undergoing significant change at this time. DOE is preparing a revised
PC plan that will supersede its earlier plan. A new "portfolio" of PC activities has been selected
using a multiattribute utility analysis. The selected portfolio is now being reviewed for approval
by DOE's management. When approved, Revision 1 of the plan will be provided to the NRC.
It is expected that a Revision 2, to be published in 2004, will include a full description of each
PC activity. The staff intends to use the review methods in the Yucca Mountain Review Plan to
perform pre-licensing reviews of Revisions 1 and 2 of DOE's PC plan.

Observation

A PC plan is required to be a part of a license application; therefore it is clear that this element
of DOE's program should receive appropriate pre-licensing guidance. Based on NRC's
presentations to the Committee, however, the PC program has not been treated proactively by
NRC. The staff is waiting for DOE to propose a structure for a PC plan and to suggest criteria
for deciding whether deviations from baseline are significant enough to warrant actions. We
believe that PC is an area that deserves more interaction between DOE and NRC than has
occurred to date.

Recommendations

The Committee recommends that the Commission require the NRC staff to provide additional pre-licensing guidance to DOE concerning PC plans. These communications should focus on:

1. Ways to develop the PC program that are based primarily on risk insights and testing assumptions about key performance factors;

2. How performance assessments can or should be updated using performance confirmation data;

3. How performance confirmation should be used in making decisions; and

4. How to resolve any differences in NRC and DOE approaches to PC.

Attributes of a Successful PC Program

The PC Program Should Be Informed by Performance Assessments

The PC program must be risk-based, focusing on parameters and processes that are important to safety. PC needs to be linked to total system performance assessments (TSPA for DOE and TPA for NRC), which means these assessments have to be maintained during PC. Also, PC monitoring should focus on areas where TSPA is based more on assumptions than on evidence. To the extent that TSPA and TPA indicate that performance is insensitive to some systems and processes, monitoring of associated parameters may not be needed. A risk-based PC program would allocate resources to those areas that are most important for performance, thus providing the greatest support for future decisions.

NRC's review of DOE's PC Plan may identify elements that are unnecessary and not risk informed. The staff normally focuses licensing reviews on activities that are needed but have not been proposed by an applicant. The NRC staff seldom comments on unnecessary activities that an applicant may propose. However, in a risk-informed, performance-based arena, it is appropriate to provide guidance to a potential licensee regarding both necessary and unnecessary activities.

To avoid the pitfall of having the PC program become a de facto site characterization or basic research program, there should be a clear mapping between performance assessment and PC.

The PC Program Should be Flexible and Responsive

Considerable advances in technology can be expected to occur over many decades. A successful PC program should be flexible, with a process to reevaluate, reexamine, and modify PC activities as the state of understanding changes. New tests may be needed, or may become possible with new technology, and tests that are no longer providing useful information could be discontinued. Some parameters are difficult to measure but nonetheless may be important to safety. The Committee advises an approach to develop and correlate new data, to the extent feasible, to build a body of evidence that will improve the safety-related knowledge base.

Objective Criteria Are Needed To Decide on Future Actions
The PC plan should address what happens if some results are unexpected and potentially at odds with assumptions used in development of the safety case. PC is not aimed at detecting performance failures per se. However, the PC program may detect parameters that deviate from an expected range of values. Yucca Mountain is a complex project, so that some deviations from expectations may occur. PC should have a logical pathway to determine whether any of the deviations are significant to safety. The criteria to make this determination should be developed as part of the PC plan.

Appropriate Accuracy or Precision Should be Part of the Measurement Design

Parameters to be monitored under PC will require varying degrees of accuracy and precision to support decisionmaking. The appropriate metric should be whether significant deviations important to safety have been detected. Requiring unnecessary accuracy or precision may be misleading regarding the importance of the parameter.

Plan Should Include Appropriate Involvement of the Public in PC Activities

The PC plan should address how the public will be involved in the PC process. The public could be involved in identifying those aspects of a PC program that may provide increased confidence. The Committee believes that the PC plan needs to be risk informed. However, some activities may be planned to address issues of unusual public concern, though they may not be high-risk safety issues. The public should be kept informed of any problems revealed by the PC process and of any subsequent mitigation.

Summary

This WGS provided an excellent forum in which to exchange views on the technical issues associated with PC. It appears to the Committee that, within the high-level waste program, PC planning is relatively immature. The Committee has provided specific recommendations to enhance the pre-licensing guidance so that DOE can improve its PC plan. NRC and DOE have not yet finalized any agreement items related to PC. Continued communication between the NRC and DOE staffs is essential, and must stay focused on matters important to safety.

Sincerely,

/S/

B. John Garrick
Chairman

Left blank intentionally.

3. KEYNOTE PRESENTATION

PERFORMANCE CONFIRMATION FOR YUCCA MOUNTAIN

PRESENTATION BY DR. CHRIS WHIPPLE
ENVIRON
JULY 29, 2003

The first slide [Slide 2, after title slide] is an overview of what I hope to cover this morning. I will cover performance confirmation in a philosophical sense. How should we think about performance confirmation and what should it be? How do we decide what testing to include and what to exclude? What activities should be done based on criteria that make sense, and what shouldn't be tried in performance confirmation?

An earlier agenda had listed presentations on WIPP and a later agenda didn't. Until Wendell walked in this morning, I didn't know that someone who knew a lot about WIPP was going to be here. Nonetheless, I think there is a lot we can learn about the process followed at WIPP that applies to performance confirmation at Yucca Mountain. Then I want to talk about some specific technical arenas and discuss whether they make sense as candidates for performance confirmation.

[Slide 3]. First comment. These are my thoughts, and DOE has not seen my slides. They haven't commented on them, obviously, if they haven't seen them. I have heard from talking to somebody in the project that Karen Jenni and Jim Blink had worked up a new performance confirmation plan for the project. Karen and I talked and we agreed it would be better if we didn't see each other's slides in advance. This talk was not intended to be a review of a document but, rather, thoughts on what performance confirmation is. So I did want to get this disclaimer in.

The second qualifier is that several years ago a group of us helped John Kessler put on a workshop at EPRI on performance confirmation. I think some of the people here took part in that. We produced the proceedings from that, and I had various notes in a talk I gave there.

I "stole" liberally from everyone's contributions to that workshop in thinking about this presentation. Some of the ideas that I "stole" were mine originally and others weren't, but I thought that was a good workshop. I recommend those proceedings if you haven't seen them.

[Slide 4]. First is a starting point. The word "confirmation" is just a lousy word. It suggests we're certain of everything and we're going to nail it down and confirm it. I understand a licensing process is a legal process, but I am a technical person. There will always be uncertainties in performance and our understanding of performance. I think it's sensible as a technical person that we continue to refine our understanding, even when we believe we have crossed the threshold that says we know enough to issue a license and begin operations. But the tone of the word "confirmation" suggests that we can't disqualify what we know. And that's really the main point of performance confirmation as I see it. You can wander into the philosophy of science literature, and you find that hypotheses are only falsifiable. You can't confirm them. You can only prove them wrong.

A major objective of performance confirmation is to look for signals that we've got it wrong and that the repository might not be appropriately safe. I think that should be the driving objective.

11

[Slide 5]. How do we go about that? One of the things that came out of the EPRI workshop was a list of desired aspects for any performance confirmation program. And a little later in the talk when I mention WIPP, you'll find that some of these management principles have been missing from the WIPP project at high cost to that program and to the public that pays for it.

It's important to understand the need to be flexible and iterative in anything we do. We need to preserve the ability to start something in performance confirmation, get a year or two in, and say, "This isn't telling us anything useful. We might as well pull the plug on it." That's hard to do in a setting in which activities are undertaken by enforceable agreements, but it really is appropriate for a program that is going to involve a fair amount of learning as we go, which I think performance confirmation will.

The term "risk-informed," of course, was invented here. I shouldn't have to preach to the choir about that. But I think NRC's regulation 10 CFR Part 63 has missed the boat on performance confirmation in some aspects. The issue for me for performance confirmation is how it connects to the high degree of safety that we desire at a repository, not verification of DOE paperwork.

Something that is difficult to do but essential is to apply performance confirmation in a manner that gives the public confidence that if the repository deviates from acceptable performance, we have a chance to identify and fix it, reverse it, do something about it. The public needs to be involved in identifying those aspects of performance confirmation that can provide increased confidence.

I mentioned "iterative" in my last slide. I think it's possible over an indefinite but long operating period, 30 to a couple of hundred years, to think in stages. Performance confirmation tests should not be specified at the time a license is issued and then applied unchanged for 200 years.

The other aspect that is terribly important is the need to base priorities on sensitivity of overall performance. That is, we have to keep our eye on the ball of "Does it matter?" And then, finally, one of the things the project deserves a lot of credit for is the ability to overcome the temptation to lock everything in 10 years ago. There have been many improvements in the design and a lot of improvements in the analysis. And I hope that exploratory mindset can be maintained over the long performance confirmation period.

In terms of our ability to analyze, model the subsurface performance, particularly unsaturated zone performance, the science there is really somewhat immature. Twenty years ago what we could do compared to today was practically nonexistent. And one hopes 20 years from now the science will be a lot better and the performance confirmation process will evolve accordingly.

[Slide 6]. 10 CFR 63.131 requires a review to see if the conditions in the subsurface are consistent with those assumed in the license application, and to see if the natural and engineered systems are performing as anticipated.

The word "safety" doesn't appear here. To me, I read this as a statement that the performance confirmation is focused on going back and retrospectively looking at whether the license application is still up to date when we are 10 or 20 years down the road and have more data from underground, and not whether we have new insights as to whether the appropriate limits for public protection are met or not.

I would have preferred that the safety emphasis had been stronger and that what I see as a consistency of paperwork was secondary to the higher level goal of protecting the public. I suspect we can talk about that over the next few days.

All right. So my second major bullet there is the question I just asked, "Are we there to confirm paperwork or to confirm safety?" The final point is, "To what extent do we want to continue to reduce uncertainties?" And do we want to do that across the board or do we want to do that only for those things that are truly significant to safety?

It is not unknown in a big, complicated project like this one to have large teams of people whose careers are involved in polishing the third decimal place. And I hope we don't do too much of that.

[Slide 7]. This slide came out of the EPRI workshop and refers to pitfalls to be avoided. I thought it was on the money then, and I still think it is. As you approach the hectic activity of assembling a license application, it's tempting to deal with a lot of problems by putting them into performance confirmation. In other words, performance confirmation can become the "bucket" into which you put the problems you can't solve this week. This can get you into trouble in a number of ways. First is the obvious one. You shouldn't agree to do anything that can't be done. It will come back and bite you in a big way, and it only postpones the pain of dealing with things.

Another point is—and I will hit this one again later—agreeing to measure things that don't matter. It's just think a generally poor idea, its expensive, and it takes attention away from things that do matter.

Third problem, and I hope this is not something that someone does, but 15 minutes into monitoring, I hope no one says, "See, the repository is safe. We don't detect any radiation whatsoever in the groundwater 20 kilometers down gradient."

Well, of course nothing would be detected in such a short time. It wouldn't prove anything about the safety of the repository. That's something we have to be very careful about, which is to monitor things that are meaningful.

One of the things I mentioned earlier is that if the public thinks it's important to do it, you do it. And I suspect monitoring groundwater where people are may well climb onto that list. And that's fine if that's what people think is important. But you shouldn't claim that because radiation hasn't shown up in 100 years, that proves the safety of anything.

Another aspect—and I'll get to this in talking about some of the WIPP stuff— is don't agree to measure things plus or minus five percent when what you really need is plus or minus two orders of magnitude. Measuring with unnecessary accuracy and precision misstates the importance of what you are trying to do, and is also more expensive.

The right starting point should not be, "How well can I measure this if I use the best available technical means?" It's "How much does this matter? And how well will I need to know it?"

Then, finally, back to that word "iterative," just because you agreed to do it at the time of the license doesn't mean that it will make sense 10, 20, or 30 years from now. And from the start you need a process to reevaluate, reexamine, add, and delete performance confirmation requirements as the state of understanding changes.

[Slide 8]. Performance confirmation in my view—and this may be tailored by having spent a lot of time looking at DOE's Total System Performance Assessment (TSPA)— will be tightly linked to TSPA. The TSPA is the core of the license application's case that compliance has been achieved. The question, then, is what can you monitor in performance confirmation that is predicted in TSPA and has a bearing on meeting the primary safety objectives.

The other point is that to continue that monitoring 30, 40, 50 years into the future implies that you are going to maintain TSPA as a living model. That "living model" term comes out of the PRAs used in the nuclear power plants. The plants tend to keep them up to date. The models evolve with time, and incorporate modifications to the plant or to our understanding of the plants.

I'm simply ignorant about whether TSPA will become a "living model" for Yucca Mountain. I know at WIPP, there is a requirement for recertification every 5 years. That has kept a certain amount of activity going on their performance assessment, but I must say I had the impression there was about a 4-year dormancy period and then "Oh, my God. We've got to get the thing recertified in a year. We had better kick this thing back to life."

I don't know what will happen with the Yucca Mountain TSPA, but if you intend to maintain a linkage between performance confirmation and your understanding of the site, the TSPA has to be kept alive.

[Slide 9]. Here is where I play the role of Karen Jenni and try to determine what decision criteria are needed for performance confirmation. I came up with four general categories, and there is a slide on each.

The first category is a simple one. It matters to safety. If we can monitor things that affect our belief about whether or not the regulatory dose limits are met, then that is an obvious one. The second category is that some parts of TSPA are oversimplified. They're bounding analyses. We know they're weak.

Anyone who has read the near-field environment section of TSPA more than twice knows there are parts of that process that we don't understand very well and we can't model very well. I don't mean just to pick on that one topic, but there are several other topics like that.

It may be useful to monitor in areas where we believe TSPA is weak. But to the extent that we think TSPA has bounded a reasonable worst case, such as where waste packages start leaking immediately, then you may not need to do it based on that first point if it doesn't matter to safety.

A third category, TSPA is loaded with any number of conceptual models. And the project team has done a lot of work to try to evaluate those conceptual models and test them against alternative conceptual models. But, again, field evidence that bears on the question "Do we have a basic correct understanding of this or that process?" could be terribly important. And then the fourth category I mentioned before is where the work would address an issue of public concern, even if it didn't meet some threshold of importance to safety.

[Slide 10]. In terms of "importance to safety," the question is "Are we on an absolute or relative scale?" By that I mean how does this affect compliance with a 10-millirem-per-year dose limit within 10,000 years? That is an absolute scale.

14

A relative scale says, does this matter more than 10 percent to the calculated doses at future times? That would say by some threshold—and I picked 10 percent out of the air—this is a relatively important factor compared to the other 189 factors in TSPA. Perhaps we should worry about it.

Either way, I think those two ways of asking the question, "Is it important to the absolute achievement of dose limits" or "Is it important to understanding the relative contributors to performance," are preferable to the question, "Is this consistent with what DOE told us in their license application, and whether or not it matters?" I will keep hammering away at this theme.

[Slide 11]. There has been a great deal of work done with limited success across the whole risk analysis field in trying to deal with the problem of alternative conceptual models.

Proposals have been made to use weighted averages of different models. That satisfies no one and simply assures that you are going to be only partially wrong, not completely wrong. Some of the related work using sensitivity studies of parameters and alternative models has been helpful in understanding the importance of relative subsystems, but there is always some concern about it because if the model is totally wrong, then you can't rely on the sensitivities either.

One of the things that I hope that can be done in a thoughtful way is to worry about where TSPA is weak and can perform its confirmation, supplement the knowledge there with the condition that things matter.

The final bullet on the slide is the qualifier "it needs to matter." Confirmation activities where TSPA is nonconservative, where meaningful measurements can be made, and where an issue is important to safety may be a small set when you finish running through those three filters. But that is the kind of thing you should be worrying about and looking for.

[Slide 12]. This slide relates strongly to the last one. It goes after the question "Can you take measurements about the relative credibility of competing conceptual models?"

In the WIPP project over the years, there was a running fight over matrix flow versus fracture flow versus dual-phase, dual-media flow. In the long run, they converged on a set of models where it didn't matter a whole lot whether you went with just fracture flow or with two-media flow. The water moved about as fast in both cases.

Starting in the late 1970s, and the first simple models of an underground repository, the basis for the first EPA [high-level waste] standard tended to begin with an assumption that rock was homogeneous. Over time we have come to understand that is not even true in an salt site like WIPP. It certainly is not true for a hard rock site like Yucca Mountain. It matters that there are fast flow pathways and we have to be aware of them. Getting the conceptual model for the fast pathways is hard.

I am not sure that performance confirmation is going to be better than what we can do with tests in the existing tunnels now. A lot of people are looking at thermal effects in performance confirmation. In the grand scheme of performance assessment, thermal effects tend to be transient and not necessarily of high importance to safety, although that can be debated.

[Slide 13]. There needs to be a category for performance confirmation because the public worries about it. If you spent some time reading the risk communication literature, probably the single most important recommendation is to talk to people about what they're worried about.

A favorite example of mine is for years polling done by the nuclear utilities showed that people worried that nuclear power plants could blow up like atomic bombs. The nuclear power industry people knew this was impossible and, therefore, not worthy of discussion. Therefore, neighbors of power plants went on worrying that these things were going to blow up like atomic bombs.

If people are worried about something that you think is unimportant, that is a great topic for conversation. And if they are worried about something where you don't think you can do meaningful measurements but they want them anyway, well, that is probably a price you have to pay.

The subtext on this has to be that you should not assume that DOE's managers understand what the public worries about and what they would like to see done. That would be a serious mistake.

I am afraid a process is needed. There is a legitimate basis to include activities in performance confirmation because they are subjects of public concern and because the action itself provides some reassurance.

It shouldn't be an excuse for some idea that couldn't meet any of the other criteria for being carried out under performance confirmation. If someone has a pet project that is completely unimportant to safety, they may argue we should do it because the public wants it. There needs to be a threshold to decide if something is important enough to safety to include in confirmation.

[Slide 14]. The U.S. has cleaned up hundreds of Superfund sites. These sites were usually on the surface, very close to where people are, and could often be fixed with much less expensive remedies than could potentially be used at Yucca Mountain.

There are processes for thinking through the continuing monitoring requirements. In the EPA world, one approach used is called the data quality objective framework. Among decision analysts, they use a term called "value of information." Both have the same key idea, which is if you are measuring something that does not affect any decision you make, then you probably shouldn't be measuring it?

The question "Has it leaked yet?" is still a fair question to ask. As long as the answer is no, you might argue that no decision is being made, but, in fact, the decision is we don't have to go back in and patch. That is a decision. I think this framework could be constructively applied in the case of Yucca Mountain.

The question is, where would measurements make a difference, either to changes in design, changes in operation, or to remediation of something, patching and fixing. What measurements could lead to a decision that we've got it all wrong and we have to retrieve waste?

There is a correlated issue here, which is that the NRC needs to worry today about what happens when performance confirmation measurements fail to track with TSPA predictions. Do you say, "That's too bad"? Do you say, "Resubmit the license"? Do you say, "Do an analysis that shows you still comply with a 10-millirem dose limit?" Those things need to be thought through.

In something as complicated as Yucca Mountain there will likely be deviations. How do you determine which are significant? Is a 10-percent difference from what I predicted in terms of the temperature profile on the rock significant, or is that a trivial difference?

All of those things need to be thought through because when you suddenly have the data, then it is harder to develop criteria that you wish you had done objectively beforehand.

[Slide 15]. Now for a few slides about the Waste Isolation Pilot Project (WIPP). When the WIPP project was at about the same place in its evolution as Yucca Mountain is today; that is, when the certification compliance application was being prepared for review by EPA, there were lots of cats and dogs that hadn't been put to bed, lots of minor technical issues still unresolved.

You might remember, there was a painful phase in the WIPP project where DOE proposed to run experiments by putting about 10 or 15 percent of the waste into WIPP ahead of its license just as an experiment. Many people, including me, saw that as simply an excuse to get people in New Mexico used to the idea that WIPP was going to open. And I didn't think it had any technical merit.

The fact is that the WIPP project when it was being considered had a lot of requirements that had to be developed. One of the most important requirements was the waste characteristic analyses to be performed.

EPA did try to do DOE a favor. EPA in their draft regulation offered DOE several choices. It basically said, "We invite DOE to come to us with a sensible plan for waste characterization, and we will review it. That plan might include statistical methods. It might include working backwards from performance assessments to determine what ranges of waste characteristics could affect a determination of compliance or any other method that DOE wants to propose. We [EPA] will be happy to review the plan."

Absent that, here are 97 pages that we xeroxed from the RCRA (Resource Conservation and Recovery Act) standard that say you have to measure absolutely everything about every piece of waste that you propose to put into WIPP. DOE did not submit a plan to EPA that time. This was in the late '80s. I remember being horrified by this and talking to the WIPP project manager. I'm paraphrasing his answer, but the answer is that last bullet. "I know we have to have that fight, but I want to have it on the other side of the finish line."

The view was that trying to negotiate all of those requirements while you're trying to get your license will delay getting a license. And it wasn't said at the time, but I think there was a sense that it gives EPA a lot of leverage over requiring things that are excessive compared to what we might do later when they don't have that leverage of do you want your license or not. What DOE misunderstood is how hard it was going to be to try to fix these excessive requirements after the fact.

[Slide 16]. On the EPA side, characterizing the radiological aspect of the WIPP waste is pretty straightforward. Radiation is easy to count. Furthermore, for the waste that goes into WIPP, the hazard is predominantly radioactive. The chemical hazard is trivial relative to the radiological hazard. Nonetheless, the bulk of the money in waste characterization at WIPP goes to characterize chemical waste.

17

Part of the reason for that is that the agreed-to waste characterization requirements, which DOE proposed to New Mexico, included enormous detail. DOE promised to measure everything. New Mexico said, "It sounds fine to us. Let's agree on it. Here's your RCRA permit."

[Slide 17]. DOE has tried to reevaluate the waste characterization requirements. It has proven difficult. New Mexico says, "Oh, wait a minute. We shook hands on this. You came to us and said, 'Here is what we think is a reasonable set of requirements for our RCRA permit. We promise to measure the following things if you give us a permit.' We shook on it."

DOE's view is "No, no, no. That was just to get the game started. And now that we are older and wiser and two managers down the road, we want to go back and renegotiate all of these requirements."

Right now the estimated price tag for characterizing the WIPP waste is about three billion dollars. Nobody thinks it makes sense who understands that waste.

To compound the lunacy, up at INEEL, where they have a large amount of waste bound for WIPP, they looked at the cost to characterize it. And they said, you know, "This is two to three thousand dollars a drum. For $1,000 a drum, we can treat it. We can open it up. We can compact it. We can make hockey pucks out of it. We can reduce the volume. We can give it better operating characteristics. And it will be cheaper." And that's what they're doing.

Now, it's only cheaper compared to the suboptimal overcharacterization that was agreed to initially. There are 40,000 drums of waste in WIPP. And they have measured the head space gases in every drum.

The average concentrations of those head space gases for any of 30 different chemicals do not exceed the allowable 8-hour workplace exposure limits under the OSHA standards, which is to say there's not much there. Nonetheless, DOE continues to measure the head spaces gas in every single drum.

Part of the problem there is that DOE has not made a good case for this chemical testing being unnecessary, hasn't put forth a statistical approach or any sort of approach. But it's not hard to imagine Yucca Mountain getting itself into the same predicament. DOE might agree to do everything under the sun in performance confirmation in order to speed the license application's process for the NRC.

And then once that happens, new management comes in at DOE and says, "We promised what? Do you know how much that costs? This is nuts." And all the other people at the table feel like they have been lied to. The time to figure it out is on this side of the finish line.

[Slide 18]. Just to elaborate on this, I can imagine that there will be awkward key technical issues (KTIs) and that one proposal for dealing with those awkward KTIs may be to say, you know, "We don't really have to figure this out today." Let me urge you to be very careful about doing that.

Final point on that slide, again— and this is one that I see biting the WIPP folks. It is built into their process, but their permits only last for 5 years. What was not built into their process was any sort of expectation that the requirements should fundamentally change. And change is reviewed by New Mexico as reneging on a promise.

[Slide 19]. Now I am going to mention some specific technical areas where it may or may not be useful to do performance confirmation. It is necessary to monitor for radiation leaks in the ventilation gases coming through the repository. However much you believe your TSPA and its statements that the waste packages won't leak, the fact is if you're not looking for leaks there, where you would have a chance of finding them, then one might argue that the whole performance confirmation program is essentially meaningless.

Another aspect—and this gets into an issue where there is slightly more technical uncertainty—is how likely are rock falls that could impede ventilation of a drift and could potentially damage the waste packages. Not only do you need to have an ability to detect where that happens, maybe by measuring something simple such as temperature or flow rate of the air from that given drift, but do you need to have a plan in place for dealing with such a situation? That's not part of performance confirmation, but it's part of a reasonable set of contingency plans that NRC and DOE need to have.

[Slide 20]. One of the areas where a huge amount of modeling has been done, where we really can't do realistic measurements without loading the repository, is the thermal hydraulic performance. How far does the boiling front move out into the rock wall if DOE goes with a hot repository design? Does the rock midpoint between the drifts stay acceptably below boiling?

Those are probably useful things to measure. But some work needs to be done to define what sort of acceptable accuracy is needed here. While I think that maintaining a below-boiling temperature in the columns between drifts is terribly important to avoid pooling of water above the drifts, whether it's 50 percent of the space or 30 percent or 70 percent may not be so important.

[Slide 21]. Here's another issue. Corrosion work has been going on largely at Lawrence Livermore for perhaps 5 years now for Alloy 22. They're testing a number of different chemical environments and trying to do things under accelerated conditions by making more chemically extreme conditions. But the predictions of the performance of Alloy 22 are that it behaved so well for so long a time that we still need to carry forward and get more data, and particularly data that can address the corrosion models and to see if those models match with lab experiments.

It would be very like OMB or the congressional staff to believe that an hour after the Yucca Mountain license is granted, all supporting analytical and laboratory work is unnecessary since the NRC said this place is safe enough to operate. Again, that gets into the difference between a legalistic and a technical mindset. My own view is that this is a set of experiments that really needs to continue.

[Slide 22]. Last slide. Another thing that is way too early to talk about, but it's something to fold into performance confirmation planning, is the question of "Can performance confirmation measurements tell us something about when it might be appropriate to close a repository?" My take is that the decision to close a repository is going to be largely driven by political factors, not technical factors. Those political factors will have to do with whether or not nuclear power comes back to life, with the future course of the weapons program and what wastes it might produce, with the disposition of plutonium from the weapons program, and whether and how that makes its way into Yucca Mountain.

And all of those things will affect the desired timing of closure. If Yucca Mountain is turned into a significant repository for weapons-grade plutonium, that might argue for earlier closure than a thermal hydraulicist might say is ideal. They might say, "We would sure like to ventilate this thing for another 50 years," but there may be overriding political reasons.

Nonetheless, I think that the question of when do we close should be viewed as both a political and a technical decision and we should look to see if the performance confirmation program can provide supporting information for that decision.

Performance Confirmation for Yucca Mountain

Presented to the Advisory Committee on Nuclear Waste

Rockville, MD
July 29, 2003

Chris Whipple

1

Overview

- ◆ Disclaimers/qualifiers
- ◆ General thoughts on performance confirmation
- ◆ Criteria by which one decides what to do or not to do
- ◆ Lessons from WIPP and their application to Yucca Mountain
- ◆ Specific thoughts about what performance confirmation might usefully include

ENVIRON

2

Qualifiers

- This presentation reflects my views on Performance Confirmation, and should not be taken to represent the viewpoint of anyone else or of any organization, including DOE. It has not been reviewed by DOE.

- Some of the material in this presentation comes from an EPRI workshop on Performance Confirmation and draws from the efforts and thinking of those who organized and participated in that event.

3

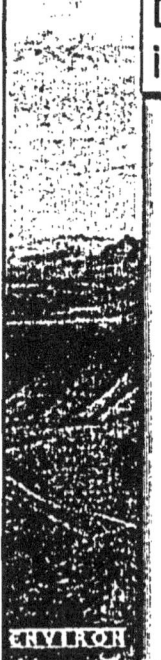

Does "Confirmation" convey the right idea?

- May indicate overconfidence
- Inconsistent with idea that hypotheses are tested by falsification
- Suggests that deviations from predictions are failures
- Deviations can indicate that the system is not as well understood as one would like, but in such cases, it is important to know whether differences reflect misspecified systems or conservative analyses

4

Management Principles

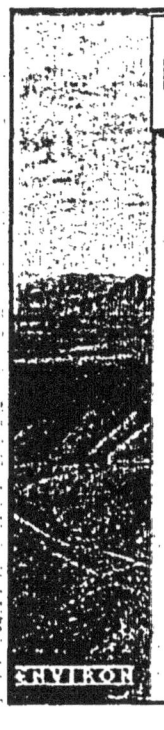

- flexible
- iterative
- risk-informed
- connected to high-level performance goals
- involves the public
- increases confidence at each stage
- can be prioritized
- has exploratory component

5

Goals for performance confirmation studies

- Part 63.131 requires performance confirmation data to assess whether
  - Actual subsurface conditions ... are within the limits assumed in the licensing review; and
  - Natural and engineered systems ... are functioning as intended and anticipated
- To what extent is such evaluation required when such conditions and systems do not bear on compliance?
- Does performance confirmation seek to reduce uncertainties in the degree of margin of performance against standards?

6

Traps

- Agreeing to do things that can't be done
- Agreeing to measure things that don't affect performance
- Claiming safety based on monitoring of too limited duration or extent
- Requiring unnecessary accuracy or precision in measurements
- Failing to establish and apply a system for periodic reconsideration of performance confirmation requirements

7

Performance Confirmation and TSPA

- Given that TSPA is the basis for licensing of Yucca Mountain, it is logical that it will also be used to determine what to monitor during the performance confirmation period.
- Will TSPA become a living model, evolving in response to performance confirmation information?
- Are periodic revisions and updates planned?
- What post-licensing level of effort, relative to current activities, is planned?

8

Criteria for Selecting Performance Confirmation Activities

- ◆ Threshold of importance based on TSPA results and sensitivity studies

- ◆ Potentially important processes or events not treated realistically in TSPA

- ◆ Can contribute to assessing the validity of an important TSPA conceptual model

- ◆ Addresses an issue of public concern, even if deemed unimportant by TSPA

9

Threshold of importance based on TSPA results and sensitivity studies

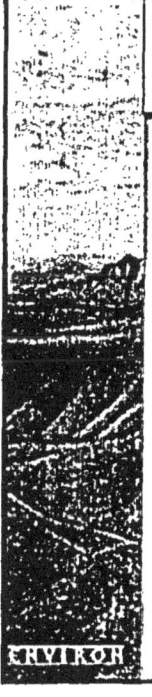

Absolute or relative scale?

- ◆ Should the threshold for undertaking a confirmation activity be that noncompliance is possible?

- ◆ Is it sufficient to require confirmation measurements for parameters or processes that are important to safety in a relative sense, but where noncompliance is not feasible?

10

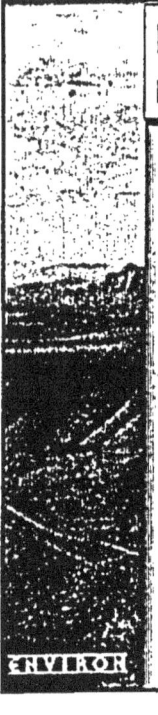

Potentially important but not treated realistically in TSPA

- There are process that TSPA treats via simplified bounding analyses, or doesn't address where the failure to do so is in the conservative direction (e.g., effect of spent fuel alteration products on radionuclide mobility).
  - Not clear where such processes can be monitored with the expectation of learning anything within the performance confirmation period
  - Not clear that it is the role of performance confirmation to make TSPA more realistic where it is conservative
  - Confirmation actions appropriate where TSPA is non-conservative AND where meaningful measurements could be made AND where the issue meets an important-to-safety threshold (may be moot given that non-conservative TSPA is probably unacceptable)

11

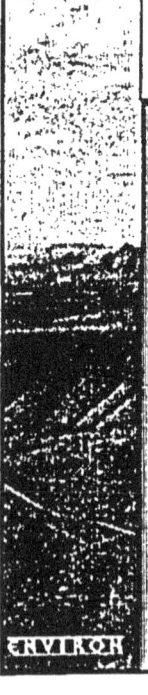

Can contribute to assessing the validity of an important TSPA conceptual model

- TSPA sensitivity analyses have been made to assess the relative importance of parameters, assuming that the overall framework is conceptually valid

- Some analyses of alternative conceptual models has also been done

- Conceptual model uncertainty is typically more difficult to address in an analysis than parameter uncertainty

- Opportunities to evaluate conceptual model uncertainties outside of the TSPA framework may be available

12

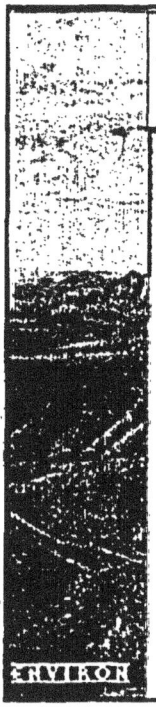

Address issues of public concern, even if deemed unimportant in TSPA

- Key risk communication recommendation is to take the public's concerns seriously and to address these concerns even if they are not seen as valid by technical experts

- Should not be used as an excuse for doing otherwise low-valued work

13

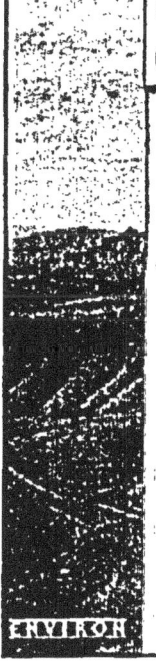

Use a value of information or data quality objective framework

- Under such a framework, data are only collected where they could affect some action or decision

- Concurrent with performance confirmation measurements, has NRC/DOE tried to define criteria that would trigger modifications to the repository or its operation? That is, how do performance confirmation data matter?

14

27

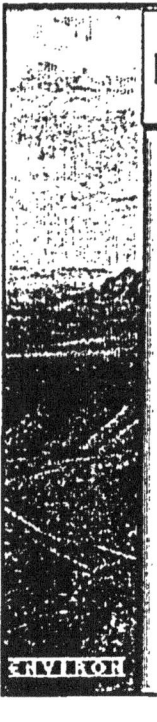

Learn from WIPP

- ◆ To speed EPA certification of WIPP's compliance, DOE deferred resolution of several key technical issues in waste characterization until after certification was granted.

- ◆ The plan was to get some waste underground, and to reopen discussion regarding characterization requirements later.

- ◆ "I know we have to have that fight, but I want to have it on the other side of the finish line."

15

Learn from WIPP, cont.

- ◆ Characterization of WIPP waste for radiological properties is managed by EPA. These requirements are straightforward; radiation is easy to measure.

- ◆ Characterization to identify hazardous chemicals is conducted under a RCRA permit granted by the New Mexico Environment Department (NMED).

- ◆ These requirements largely reflect methods proposed by DOE in its permit application. The requirements are excessive, given the comparatively minor chemical hazard of the waste.

16

Learn from WIPP, cont.

- ◆ NMED views the agreed-to permit requirements as something that DOE and New Mexico shook hands on, not as a temporary set of requirements to be renegotiated at the first opportunity.

- ◆ When WIPP opened, the budget for analysis was cut to essentially nothing. The view at OMB and among Congressional staff was that if EPA had certified WIPP as safe to operate, no significant technical uncertainties remained. Needed analyses to support reduced waste characterization have not been performed.

17

Applying lessons to Yucca Mountain

- ◆ Do not use performance confirmation as a way to put off dealing with awkward KTIs, except when it makes sense, i.e., when informative measurements can be made AND where the issue is important to safety/ compliance.

- ◆ It is normal for technical people to think their issue is the most important issue, and that it deserves a prominent place in performance confirmation – they all can't be right about this. Also need to beware of rice bowls.

- ◆ Plan for the periodic review of requirements with the expectation that they should change as data become available.

18

Monitoring to address conditions during the confirmation period

- Is monitoring of ventilation gases for radionuclides sufficient to detect early waste package failures? Other environmental monitoring, e.g., of ground water, is likely to be useless, but may provide public confidence.

- Rockfalls, while not anticipated in the confirmation period, could affect ventilation and thermal conditions beyond those analyzed in TSPA. Would monitoring of ventilation flow rates be sufficient to identify if rockfalls have occurred?

19

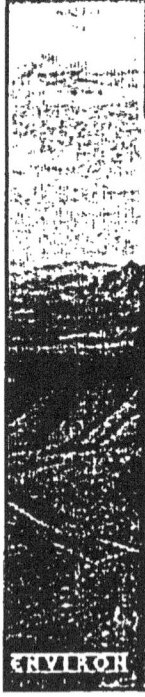

Thermal hydrologic predictions could be tested

- It should be possible to monitor and compare temperature and moisture conditions with model predictions.

- Below-boiling temperature in pillars between drifts is important to allow drainage, but peak temperatures are not reached until after closure.

- Compliance and long-term performance are insensitive to such factors in TSPA. It isn't clear how this information would be used or whether it would be informative with respect to safety.

20

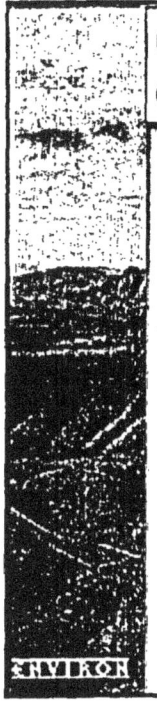

Corrosion modeling is based on limited experimental evidence

- ◆ Value in continuing corrosion experiments in a way that addresses both models and parameters

- ◆ The chemical environment on waste package surfaces will change after repository closure. It may not be possible to make measurements during the operating period that provide useful information with respect to these environments

21

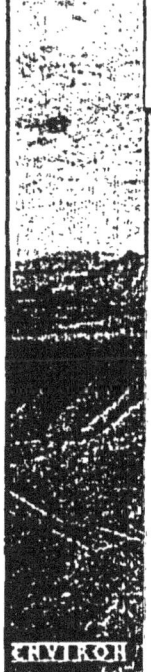

When to close the repository?

- ◆ Are there confirmation measurements that can help inform this decision?

- ◆ Some decision factors will likely involve the future course of nuclear power and the weapons program; these are not connected to confirmation.

- ◆ Current NRC requirements do not envision a post-closure confirmation program. Can useful post-closure measurements be made? Post-closure monitoring assumed for hazardous surface facilities.

22

Left blank intentionally.

MEMBER HORNBERGER: Chris, you outlined the WIPP example where DOE basically signed in to do too much and fell into one of your traps in your earlier slides. I know you have had a lot of experience with DOE. And, as you pointed out, there is lots of other experience. If you do a rough calculation in your head of things like the agreements made at Hanford and other places for cleanup, can you give us an idea of what fraction of the time DOE actually got it right so that we have some sense of the probability of them getting it right at Yucca Mountain?

DR. WHIPPLE: Well, "getting it right" is not the right term, George. I'll say why. DOE in the end usually gets it right, but it took longer and more money than it might have taken if somebody were doing it who wasn't doing it with public funds. Regarding the other point, given the size and insularity of the DOE programs, I don't know whether they learn as much from experience as they should. Certainly at the sites there has been a lot of progress. Hanford went from being a plutonium production facility to an environmental project in a relatively short time. DOE didn't change the people doing the work. It took a lot of time for that group of people to learn the new rules.

DOE is still slowly learning how to be externally regulated. They're not particularly good at it. They fight like hell over trivia. They roll over and play dead on the expensive stuff. That's not how a smart private firm is regulated. A smart private firm says, "We'll give the regulators all the cheap stuff they ask for, whether it matters or not, and we'll fall on our sword over the two things that cost all the money in the world that we think aren't really required." I don't see DOE being good about that yet.

I don't see as much of the site cleanup work as I used to. My impression is that they are getting better at that. They do have some early closure success stories now. Particularly Rocky Flats is held up as an example where I think the contractor has done a good job of telling DOE, "You have given us performance milestones, award fees based on achievements of the milestones. You don't get to tell us how to do the details because if we do it your way, we can't get it done."

I will repeat a funny old story. Back when Leo Duffy was running DOE's EM (Office of Environmental Management) and this was when the budget for DOE's site cleanups went from half a billion to five billion in a short period, Leo was in his confirmation hearing to be appointed to that job at DOE. He was coming out of running waste management services for Westinghouse.

Some member of Congress had been handed a set of tough questions. They wrote the line, "Mr. Duffy, isn't it true that when Westinghouse Electric Corporation does cleanup work for private clients, it doesn't require the full indemnification that Westinghouse requires of DOE?" And Duffy said, "Yes, Congressman. That's exactly right." The congressman kind of grinned. You know, I think he's thinking, "I've got him." He says, you know, "Do you think that's fair to the taxpayer?" And Leo said, "Congressman, Westinghouse—I'll go on record here—would be delighted to work for DOE on the same terms we work for our private clients." And the congressman knew he had been had at this point and had to say, "Oh? What's that?" Leo said, "First, we charge our commercial fees. And second, we don't let the client tell us how to do our jobs."

I think that is a problem with DOE. They hire good people but they override them at times. And, as I say, they're still learning how to be regulated externally.

MEMBER LEVENSON: Chris, you've been involved in this a long time and attended a lot of meetings. Anywhere along the line, is the issue of confirmation as an adder-on to decisions made by other people the wrong way to do it?

For instance, just one example off the top of my head is, rather than trying to monitor container failure by radioactive gases, which on very old fuel there isn't much of anyway, you might put an inert tracer in waste containers and monitor ventilation systems for that.

The basic concept of can you improve confirmation by something you do in the active program, have you seen that concept in your background or experience?

DR. WHIPPLE: Not much, Milt. Back in the late 1980s we had this terrific old chemist on the WIPP committee who wanted to put a durable blue dye in the repository. If you found it in a well you would wonder, "What on earth is this? And how did it get there?" No one took that idea seriously. And I must say I don't know of anywhere where that is being done.

I do think that these materials serve as their own tracers pretty well most of the time. But what you're asking does pose the question of integrating across discrete boundaries in the project.

I just finished service on an Academy of Sciences panel that was terminated prematurely by DOE. It was on long-term stewardship of DOE sites. The key message from that committee—we finished the report anyway—was that DOE needs to think about how it is going to do long-term stewardship of the sites as it plans the site closure remedy. DOE took great offense and said, "Yes, we do that, but we can't show you where we have written it down ever."

So I do think that that kind of long-term integration, including the design, is something that has some possibilities.

MEMBER LEVENSON: For instance, a tracer gas might give you data on waste package failure, at least a couple of decades earlier than looking for radioactive tracers.

DR. WHIPPLE: Yes, it could, particularly if you had waste package failures without fuel failure. Yes, you would pick up the container gas.

MEMBER LEVENSON: I think it is always that way because there is no mechanism for fuel failure until after waste package failure.
DR. WHIPPLE: You're right.

CHAIRMAN GARRICK: Chris, I think we would certainly agree that the focus for performance confirmation ought to be on those things that are important to safety. You analyze and test and monitor that.

I don't get the feeling that that is necessarily what is behind the plan that is being discussed by DOE at this time, even though in the preamble to the planning, they do say that the performance assessment will be the driving document.

My real question is the dilemma that we seem to have here that, on the one hand, we keep talking about focus and using the information and the tools we have that have been explicitly designed to provide focus, such as the performance assessment.

On the other hand, when I read the list of things that they're considering analyzing, testing, and monitoring, it's an extremely long list. And I don't get the sense that it has been mapped at the level of detail of the list to the performance assessment in any systematic and concrete way.

The other point that I am concerned about is you mentioned public involvement. To be sure, that has got to take place. But it should take place early, sooner, rather than later. It seems to me having it take place at the performance confirmation level is much too late to ever have any hope of achieving any kind of a program that has real focus to it.

Why shouldn't the strategy be one of getting public involvement in the methods that are being employed to define the program such that it is addressing issues important to safety? In other words, why wouldn't we want the public involvement up front. If it is done later on, that could just create an unmanageable situation?

DR. WHIPPLE: Well, I can see some practical difficulties. Nevada is by no means convinced that Yucca Mountain is going to be licensed, built, and operated. I can well imagine they would not be eager to assist in that process. In fact, they're suing to try to prevent it. Second, if we do the processes right, I am not sure everything has to be nailed down at the time a license application is reviewed and acted on. We have got a decade between then and between arrival of waste. And even then, if certain parts of the performance confirmation were 5 years in coming, I'm not sure that that is a fatal disqualifier. I think if you did it right with a flexible and iterative process, it in some ways would be more desirable.

Back to DOE's long list of things that are in their performance confirmation plan. I decided not to read it because what I did not want to do this morning was comment on it. But I think part of the solution there needs to be some process within the project in which there is a clear set of criteria applied to this list. Then a studious, skeptical bunch of tightwads needs to say to proponents of various tests, "Tell me again why you think this test qualifies for performance confirmation?"

In the end, it's going to be a negotiation between DOE and NRC, but my sense from looking at past DOE documents is I share your sense that DOE will sign up for far more than is necessary on the grounds that right now it's got a lot of issues with NRC and would like to solve as many of them as it can. This is a possible mechanism for doing that.

Maybe when we hear from Jim Blink and from Karen we will get a different perspective. I shouldn't speak for them.

DR. KESSLER: Chris, I certainly agree with your traps. You said don't agree to measure something that is not important, and measure only those things that are important. Yet you also said don't agree to measure things you can't measure. What, if anything, should DOE and NRC agree to do in the cases of things you cannot measure? Yet they're important.

DR. WHIPPLE: Well, I think it's unclear now whether you can make measurements of the critical metals that will confirm or refute the corrosion models, but I think it is important to keep trying. So that may be something that you can't measure at this time.

I will give you a related example of something that might be useful to measure. As Joe Payer, who knows all about the corrosion stuff better than most of us, keeps saying, the uncertainty in corrosion is the uncertainty in the environment.

We know what the nettle is. Might it be possible 5 years into operation to go in and send the robot in to get dust swipes off the waste canisters? Might that tell you something?

It doesn't tell you about the post-closure conditions, but it tells you what the starting point and the mixture of dust is and whether it's in any way different from the normal desert dust with a little bit of ground-up Yucca Mountain rock thrown in. That might be something that would reduce uncertainties. That would be kind of a creative performance confirmation idea worth doing.

MR. BERNERO: Chris, I agree with most of the comments that you brought up about the WIPP project. One thing I was wondering about is the subject of contentious scientific issues.

They may or may not be important to performance assessment, as modeled in TSPA. The public may not really be involved in some of them, but they are legitimate scientific concerns that the technical community has debated about.

Do you think these are a valid ground for doing performance confirmation measurements or would you rule them out simply because they may not affect long-term performance?

DR. WHIPPLE: I guess I would have to have a more specific situation to know. I'll give a generalization. I think management prematurely saying, "Okay, knock it off, we've decided that theory A is correct and theory B is nonsense," is a pure recipe for disaster in an agency. In general, it's best to let bad ideas die a deserved death at the hands of good science.

That is something I think each organization needs to have some freedom to deal with. However, I also think there are issues that have outlived their reasonable lifetimes, either on the grounds that it doesn't matter anyway or we have done this review 11 times.

In the case of Yucca Mountain, I think the stuff Jerry Szymanski was arguing was one that got reviewed to death. I think it has finally gone away, as far as I know.

It was long and painful, but I also think in the end the amount of work that was done helps give people confidence that this just wasn't buried by political muscle. I think that DOE's willingness to fund the most recent work at UNLV, in particular, was a very helpful step in establishing whether Szymanski was right or wrong.

MR. FRISHMAN: First of all, I'm surprised at the bait that you threw out there. You talk in your discussion about traps, that you don't see that performance confirmation should, as you put it, be the bucket for problems that couldn't be solved earlier. But at the same time, when you talk about management principles, you are looking for an exploratory component.

It seems to me that there is a line that is necessary between characterization work that should have been done versus the exploratory component in the example that you gave. An example is that the science of the unsaturated zone is still very early.

So how do we and especially the NRC's review staff figure out what the difference is between the exploratory element, as you call it, of performance confirmation and work that actually should have been done in order to gain enough confidence by the decisionmakers in a decision on reasonable expectation?

DR. WHIPPLE: Good question and a fair one that I think the NRC is going to have to deal with.

MR. FRISHMAN: I am asking you to deal with it right now.

DR. WHIPPLE: Okay. And I will try. I think there are a couple of standards you can apply. One is how well the work done to date measures up against the prevailing standards of good science in that arena.

don't think it's reasonable in any arena to say, "Let's wait until 2050 because, undoubtedly, the science will be better then." That's not a fair answer.
So has the work that has been done been of credible technical content weighed against prevailing good science standards? Second, has the uncertainty analysis been done in a similar way? And what does it show?

We may not need to understand the system perfectly. In the case of the unsaturated zone, I think that there are parts of it that are more important than others.

But I guess the other question I have is characterization absent an operating repository can only go so far. The key questions on unsaturated zone performance, the interesting ones, are where does the water go when there are hot waste cans inside? And how long does it stay away? What does it look like when it comes back? And what is the flow field around the drifts and so forth?

am not sure those are things that can be done in characterization.

DR. PARIZEK: Chris, you mentioned a lot of frustration with trying to reduce the monitoring responsibilities at WIPP. DOE was caught up with the agreements they made earlier.

Are there any examples of things you would add because you wanted the flexibility? And so would you add some monitoring or some observations that were not included at WIPP before, based on new understanding the science and engineering performance of that facility? And that would also obviously apply to Yucca Mountain by analog.

DR. WHIPPLE: Yes. At WIPP I can't think of any, actually. Waste is so thoroughly characterized that I can't think of a property left unexamined.

DR. PARIZEK: Let me bring up an example from the early discussion about gas and re-saturation. You could imagine waste, which could overpressurize the fluids and cause movement.

So is there monitoring being done of, say, gas pressure buildup, in the back-filled salt after you've backfilled? Again, these are kind of testing ideas that were troublesome at the time.

MEMBER LEVENSON: There is one, Chris. The previous National Academy committee to the one you're currently on made a recommendation. DOE had not planned to monitor effluents from oil and gas drilling in the area to get a background radiation picture before waste was put into WIPP so that you would know if you started seeing things whether or not it came from WIPP and it was an Academy committee recommendation that DOE expand that program. So there have been additions.

DR. WHIPPLE: I can think of one, Dick. And it's a replacement recommendation, which is in lieu of measuring every drum, why not just monitor the mine for volatile organics? It's a substitute. It's cheaper.

DR. PARIZEK: And that serves the same purpose?

DR. WHIPPLE: That's right.

DR. PARIZEK: That's a little bit different than some of these other monitoring issues.

DR. WHIPPLE: Right.

DR. PARIZEK: Thank you.

MEMBER RYAN: Chris, thanks for giving us a great start. You have given us a lot of food for thought, in terms of traps to think about, accuracy and precision, and lots of detail. So, really, thank you for giving us a great start. We'll look forward to your continued participation the next couple of days.

4. PERFORMANCE CONFIRMATION
(NRC'S EXPECTATIONS REGARDING CONTENT OF PERFORMANCE CONFIRMATION PLANS IN A LICENSE APPLICATION)

4.1 Performance Confirmation Program - Subpart F of 10 CFR Part 63
Jeffrey Pohle, U. S. Nuclear Regulatory Commission

Jeffrey Pohle's talk was a brief summary of requirements under NRC's site-specific regulation for Yucca Mountain. Pohle reviewed general requirements for performance confirmation, including confirmation of geotechnical and design parameters, design testing, monitoring and testing of waste packages, and other requirements. Performance confirmation must start during site characterization and continue until permanent closure.

Left blank intentionally.

PERFORMANCE CONFIRMATION PROGRAM
SUBPART F OF 10 CFR PART 63

144th Meeting of
Advisory Committee on Nuclear Waste
July 29-31, 2003

Jeffrey Pohle 301-415-6703 jap2@nrc.gov
Division of Waste Management
U.S. Nuclear Regulatory Commission

Discussion Topics

➤ General Requirements for Performance Confirmation
➤ Confirmation of Geotechnical and Design Parameters
➤ Design Testing
➤ Monitoring and Testing Waste Packages
➤ Other Relevant Requirements

General Requirements
Objective

§ 63.131(a)

Provide data, where practicable, to:

- Indicate whether actual subsurface conditions are within limits assumed in licensing review, and

- Indicate whether natural and engineered barriers are functioning as intended and anticipated

General Requirements
Program Duration

§ 63.131(b)

Program must have been started during site characterization, and it will continue until permanent closure.

General Requirements
Testing

§ 63.131(c)

Program must include in situ monitoring, laboratory and field testing, and in situ experiments, as may be appropriate to provide the data required.

General Requirements
Implementation

§ 63.131(d)

➤ Does not adversely affect the ability of the geologic and engineered elements of the repository to meet performance objectives

➤ Provides baseline information on those parameters and processes pertaining to geologic setting that may be changed by characterization, construction and operation

➤ Monitors changes from baseline of parameters that could affect repository performance

Confirmation of Geotechnical and Design Parameters

§ 63.132(a), (b), and (c)

➢ During construction and operation, continuing program of activities to confirm geotechnical and design parameters and ensure the Commission is informed if design changes needed to accommodate conditions found.

➢ Monitor subsurface conditions against design assumptions

➢ DOE identifies specific parameters and interactions between natural and engineered systems and components in Performance Confirmation Plan

Confirmation of Geotechnical and Design Parameters

§ 63.132(d) & (e)

➢ Data compared with design bases and assumptions. If significant differences, DOE determines need to modify design or construction methods and reports any changes to NRC

➢ In situ monitoring of thermomechanical response conducted until permanent closure

Design Testing

§ 63.133(a), (b), (c), and (d)

➤ Tests of engineered systems and components, as well as the thermal interaction effects of the engineered systems and components, rock, and water, must be conducted.

➤ Testing initiated as early as practicable

➤ If backfill included, must test to evaluate effectiveness of placement and compaction procedures before permanent placement begun

➤ Must test to evaluate effectiveness of seals before full-scale sealing operation begins.

Monitoring and Testing Waste Packages

§ 63.134(a), (b), (c), and (d)

➤ A program must be established at the GROA for monitoring the condition of the waste packages. Waste packages representative of those to be emplaced.

➤ Consistent with safe operations, testing environment representative of emplacement environment.

➤ Program must include laboratory experiments that focus on internal condition of waste packages. To extent practical, duplicate emplacement environment in lab.

➤ Monitoring must continue as long as practical up to the time of permanent closure.

Other Relevant Requirements

DOE's Performance Confirmation Program is subject to:

- Requirements for records and reports (§ 63.71)

- Requirements for reports of deficiencies (§ 63.73)

- Requirements for tests (§ 63.74)

- Inspection after the LA for CA is submitted (§ 63.75)

- Quality Assurance (Part 63, Subpart G)

July 29, 2003

I.2 Overview of Performance Confirmation
Deborah Barr, U. S. Department of Energy

Deborah Barr described how performance confirmation focuses on activities designed to confirm the technical basis for a licensing decision on Yucca Mountain. The program should demonstrate that the system and barriers are operating as predicted. DOE has updated its Performance Confirmation Plan for the following reasons:

- to address the requirements in 10 CFR Part 63
- to focus on the barriers that are important to waste isolation
- to use a risk-informed, performance-based approach to determine how to confirm each barrier's performance
- to ensure that the program is consistent and compatible with repository operations

In DOE's new vision of performance confirmation, a risk-informed and performance-based approach will be used to determine the complexity, extent, and number of activities to include for testing the effect of a parameter on total system performance or on a particular barrier. The program is designed to confirm operations rather than impose substantial design requirements, and it is intended to support an eventual license amendment for repository closure.

DOE has conducted a formal multiattribute utility analysis to determine the relative value of proposed performance confirmation activities. This analysis combined technical judgments with management "value" judgments on the importance of different goals. This multiattribute utility analysis is currently undergoing DOE review and is intended to be released this calendar year in Revision 2 of the Performance Confirmation Plan. Revision 3 of the plan is scheduled for release in the spring of 2004. Revision 3 will:

- define activities
- provide a "crosswalk" to current and previous testing
- establish the expected baseline for performance confirmation activities
- describe management and administration of the program
- identify needed test plans
- define the process for reporting variances from baseline and describe the appropriate corrective actions

Note: As of August 2004, neither Revision 2 nor Revision 3 of the Performance Confirmation Plan had been released by DOE].

Left blank intentionally.

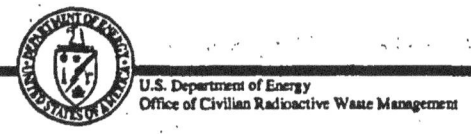 

U.S. Department of Energy
Office of Civilian Radioactive Waste Management

Overview of Performance Confirmation

Presented to:
Advisory Committee on Nuclear Waste

Presented by:
Deborah Barr
Office of Repository Development
Office of License Application and Strategy
U.S. Department of Energy

July 29, 2005
Washington, D.C.

Outline of Talks

- **Vision of the Program**

- **Focus of the** *Performance Confirmation Plan* **Revision 02**

  } D. Barr

- **Process used to select activities for inclusion into the program** } K. Jenni

- **Brief description of the selected program and its key components** } J. Blink

- **Further development of the performance confirmation program** } D. Barr

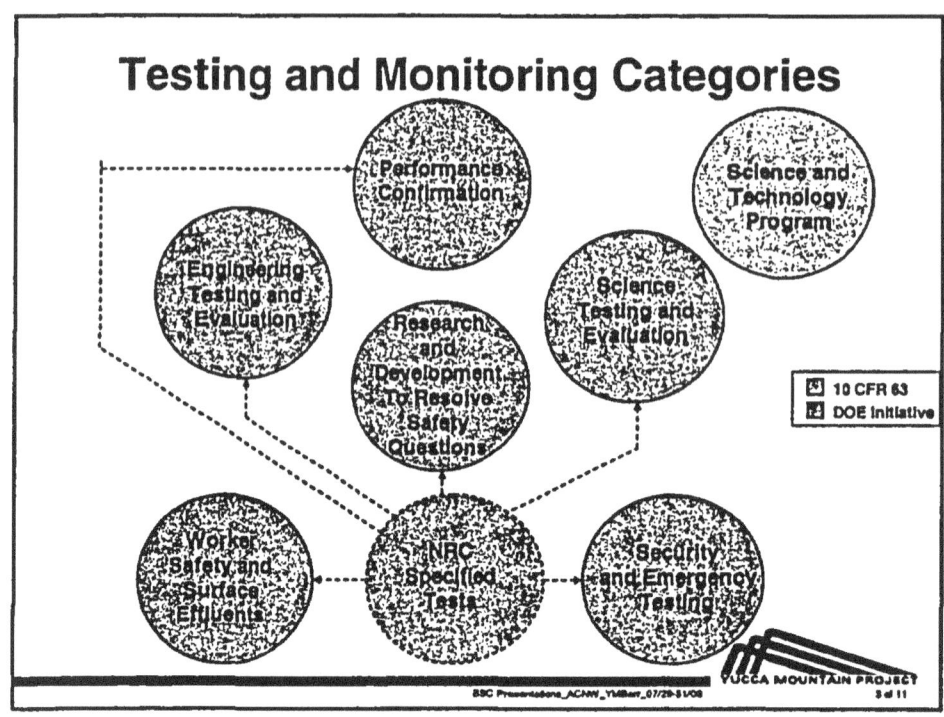

Testing and Monitoring Categories

Performance Confirmation versus Other Testing and Monitoring Programs

- Performance confirmation program focuses on
  - Activities specifically designed to confirm the technical basis for the licensing decision
  - Testing the functionality of the barriers and total system performance
- Other testing and monitoring programs focus on
  - Increasing confidence
  - Meeting other regulatory requirements

Role and Requirements for Performance Confirmation

- The NRC requires a performance confirmation plan as part of a License Application for the Yucca Mountain repository

  - "Performance confirmation means the program of tests, experiments, and analyses that is conducted to evaluate the adequacy of the information used to demonstrate compliance with the performance objectives ..." (10 CFR 63.2)

- Performance confirmation program should demonstrate that the system and the sub-system components (i.e., barriers) are operating as predicted

  - "The performance confirmation program must provide data that indicate, where practicable, whether natural and engineered systems and components required for repository operation, and that are designed or assumed to operate as barriers after permanent closure, are functioning as intended and anticipated" (10 CFR 63.131(a)(2))

YUCCA MOUNTAIN PROJECT

BSC Presentations_ACNW_YMBarr_07/29-31/03   5 of 11

Motivation to Update the Performance Confirmation Plan

- Address requirements in the finalized 10 CFR 63

  - Also address expectations laid out in the *Yucca Mountain Review Plan*

- Reflect the barriers important to waste isolation

  - Previous *Performance Confirmation Plan* based on principal factors

- Use a risk-informed performance-based process to determine how to confirm each barrier's performance

- Ensure performance confirmation program is consistent and compatible with repository operations

YUCCA MOUNTAIN PROJECT

BSC Presentations_ACNW_YMBarr_07/29-31/03   6 of 11

Elements of a Performance Confirmation Vision

- Based on 10 CFR 63 requirements and *Yucca Mountain Review Plan* expectations

- Provides a comprehensive and thorough look at critical aspects of the overall system and the barriers

- Uses a risk-informed performance-based approach to determine the complexity, extent, and number of activities to include for testing a parameter's effect on total system performance or a particular barrier functionality

- Confirms operations rather than imposing substantial design requirements (i.e., does not drive facility design)

- Supports a License Amendment for closure

Performance Confirmation Activity Selection Process

- Implemented a risk-informed performance based approach using a formal multi-attribute utility analysis of the value of including each activity

- Multi-attribute utility analysis is a decision analysis tool: used here to combine technical judgments about activities with management value judgments on the importance of different goals

Decision Analysis Based on Performance Assessment

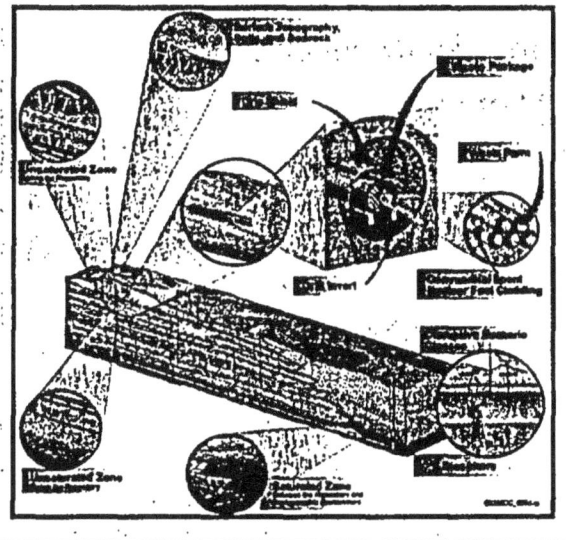

- Performance assessment barriers and scenario classes were the basis of the decision analysis

- Performance assessment technical staff provided technical judgments

- Performance assessment manager provided management value judgments

- Performance assessment includes process abstraction and total system model

---

Path Forward

- Revision 2 of the *Performance Confirmation Plan* is currently in U.S. Department of Energy review

- Revision 3 of the *Performance Confirmation Plan* is scheduled for spring of 2004

  - Define activities (what, when, where, and how)

  - Crosswalk to current and previous testing

  - Establish expected baseline for performance confirmation activities

  - Establish bounds and tolerances for key parameters

  - Management and administration

  - Identify needed test plans

  - Define the process for reporting variances and describe the appropriate corrective actions steps

Path Forward
(Continued)

- **Implement *Performance Confirmation Plan***

  - Monitor, test, and collect data

  - Analyze and evaluate data

  - Take corrective actions should significant variances arise

I.3 Decision Analysis Process Used To Develop a Performance Confirmation Program
Karen Jenni, Bechtel SAIC Company

Karen Jenni described how DOE selected the "portfolio" of tests that will constitute the performance confirmation program. DOE used a formal multi-attribute utility analysis to provide a consistent, logical, and defensible basis to compare activities being considered for inclusion in the program. Three criteria were developed to evaluate activities (and measured parameters) being considered for inclusion in a performance confirmation program:

- barrier capability and system performance sensitivity to the parameter
- confidence in the current understanding of the parameter
- accuracy with which the proposed activity measures or estimates the parameter

Technical judgments about sensitivity, confidence, and accuracy were made by the technical experts who were most familiar with the topics. A "core" team of technical experts independently assigned "utility scores" as a consistency check. Where large differences existed between the scores of the technical experts and the "core" team, the scores were discussed and reconciled until differences were small. The few differences that could not be resolved through discussions were reviewed and resolved by a knowledgeable senior manager. Costs of various activities were also considered in developing test portfolios. DOE initially received 237 parameters and 360 activities for possible inclusion in portfolios.

Altogether, DOE developed 11 portfolios, of which 6 were evaluated in detail. The portfolio designated C has been selected by DOE's BSC (Bechtel SAIC Company) Manager of Projects and senior advisors as a starting point for the performance confirmation program. This program was considered to be cost-effective and captures 82 percent of the "total potential utility." Portfolio C underwent further review by BSC senior management. Of the original 99 activities, 26 were removed because they were more logical candidates for other testing programs, 3 activities were combined with other activities in the program, 3 were retained in principle but modified in scope, and 2 new activities were added. BSC then proposed the resulting modified portfolio to DOE.

Left blank intentionally.

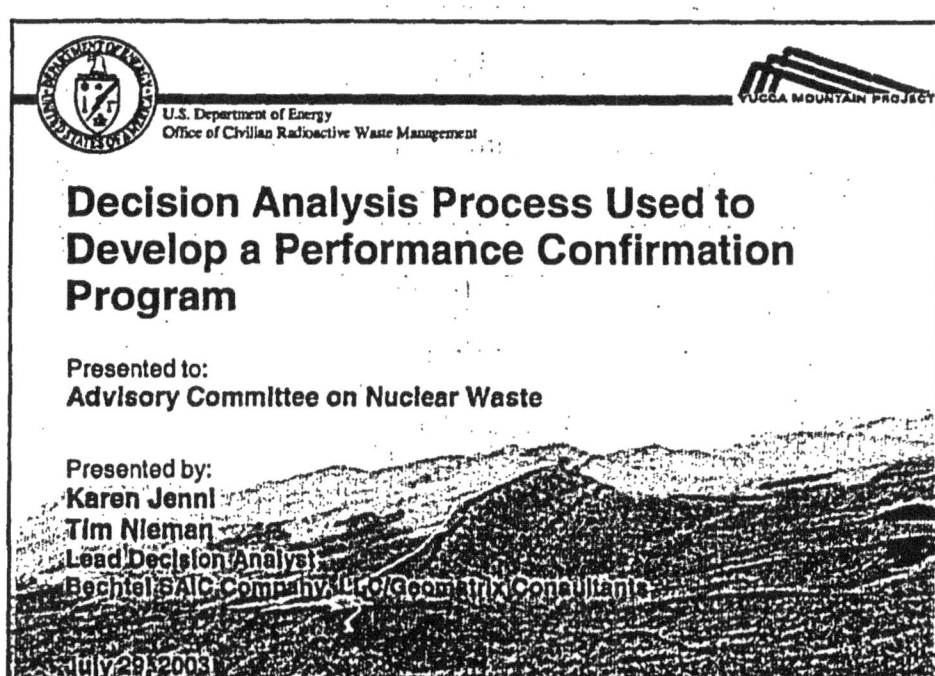

U.S. Department of Energy
Office of Civilian Radioactive Waste Management

Decision Analysis Process Used to Develop a Performance Confirmation Program

Presented to:
Advisory Committee on Nuclear Waste

Presented by:
Karen Jenni
Tim Nieman
Lead Decision Analyst
Bechtel SAIC Company LLC/Geomatrix Consultants

July 29, 2003
Washington, D.C.

The Decision Analysis Approach Separates Parameter from Portfolio Evaluation

- The performance confirmation program consists of a "portfolio" of activities
  - A set of specific activities designed to monitor or test performance confirmation parameters
- The best portfolio does not necessarily result from simply including the top ranked activities
  - There may be objectives or goals for a performance confirmation program that are unrelated to the specific activities included
  - There can be interactions among activities that make it more or less desirable to include two specific activities together
- However, the value of the portfolio depends at least in part on the value of the specific components of that portfolio
- Evaluating the individual activities is a prerequisite to evaluation of portfolios

Terminology

- *Parameters* are "things that can be measured or observed"
- *Data acquisition methods* are the means to measure parameter(s)

Parameter	Data acquisition method
Temperature and relative humidity of the waste packages	Monitor temperature and relative humidity of the air in the emplacement drifts
Temperature and relative humidity of the waste packages	Use a remotely operated vehicle to take physical measurements on the waste package surface in the emplacement drifts
Composition of the drift invert materials	Testing of invert material in the drifts prior to emplacement of waste

- Each combination of a parameter and data acquisition method is a *performance confirmation activity*
- A *portfolio* is a complete set of performance confirmation activities which could form the basis for the performance confirmation program
- The *performance confirmation program* is the selected set of performance confirmation activities

---

Decision Analysis Approach

- Provides a consistent, logical, defensible basis for evaluating and comparing activities considered for inclusion in the performance confirmation program
- Explicitly acknowledges that tradeoffs among different objectives and goals may be necessary
- Bases the evaluation on:
  - The potential impacts of including the parameter on the key objectives of the program ("technical judgments")
  - The relative importance and value of achieving those objectives ("management value judgments")
  - Combining technical judgments and management value judgments yields a "utility," or overall estimate of the value of including the potential activity
- Facilitates documentation of the technical and management basis for the selected portfolio of activities

The Technical Basis for the Approach is Formal Multi-Attribute Utility Analysis

- A technically sound mathematical approach for evaluating alternatives where more than one objective is important

- Has been used by DOE, other federal agencies, and private companies since the late 1970s to evaluate complex decision problems

- The five-step process for implementing multi-attribute utility analysis:

    - Define the objectives of the decision-maker(s), and develop metrics to measure performance against those objectives

    - Evaluate how each alternative performs against each objective

    - Assess tradeoffs: value functions and weights

    - Combine value functions and technical evaluation to estimate the overall value of each alternative

    - Use the combined evaluation results to support decision making (consider the appropriate decision rule, the quality of information, the comprehensiveness of the analysis, etc)

Approach

| Phase 1: Activity evaluation | Phase 2: Portfolio development and evaluation | Phase 3: Portfolio selection and refinement |

Management value judgments   Technical judgments

Define activity evaluation criteria → Define and describe candidate performance confirmation activities

Assign management value judgments to criteria   Evaluate activities (technical judgments using evaluation criteria)

Combine technical activity evaluation and management value judgments to get overall utility for each candidate activity

In each phase all scenario classes and barriers were explicitly considered

Activity Evaluation Criteria

- **At an initial workshop (August 26, 2002), three criteria were defined, to be used in estimating the potential impact of a performance confirmation activity on the performance confirmation program:**

  - **Barrier capability and system performance sensitivity to the parameter**

  - **Confidence in the current representation of the parameter**

  - **Accuracy with which the proposed activity measures or estimates the parameter**

- **Workshop participants included:**

  - **Technical investigators with various areas of expertise**

  - **Performance assessment analysts and managers**

  - **DOE staff**

Estimating the Utility of a Specific Activity

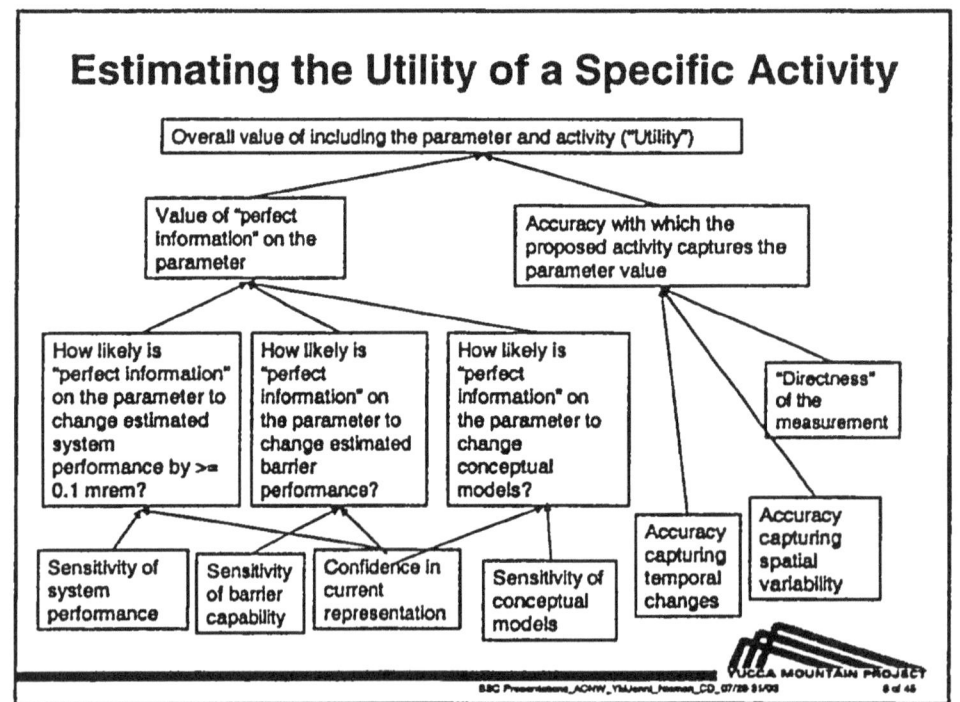

60

A Detailed Set of Questions was Developed Around Each Criterion

- The goal of the questionnaire was to elicit technical input on how well proposed parameters and activities meet the three criteria
  - Detailed questions and "scales" are also necessary to allow managerial value judgmnts to be applied consistently to the technical judgments

- The goal of the questionnaire was to improve consistency across model areas
  - Technical judgments about sensitivity, confidence, and accuracy must be made by the relevant technical experts most familiar with the model areas
  - Unaided or ad hoc evaluation of parameters by different individuals typically results in vastly different interpretations of the criteria
  - A single consistent set of questions reduces inter-individual variations in interpretation

Workshops Were Held to Develop Candidate Activities and Distribute the Questionnaire

Technical Judgments

- Workshops were held in September 2002 with each group of technical experts
  - Technical investigators and Total System Performance Assessment modelers familiar with each barrier, with total system evaluations, and with disruptive events analyses

- During the workshops
  - Each group developed a comprehensive list of parameters to be considered
  - For each parameter identified, the group defined one or more data acquisition methods that could be implemented to provide information on that parameter
  - Several activities were evaluated in each workshop by the group, using the questionnaire

Parameters were Evaluated in Small Group Meetings

- After the workshops (October-December 2002)

  - The technical experts used the questionnaire to specify their technical judgments on each activity within their area of expertise

  - A subset of the core team specified their technical judgments on each proposed activity across all model areas, to provide a consistency check

- Differences in the technical judgments by the two groups were identified and then reconciled

  - When differences in "utility scores" calculated from the evaluations differed significantly, individual scores were discussed and reconciled until the differences in the evaluations were relatively small

    - "Significant" differences in utility were defined as differences larger than 10 percent of the difference in score between the highest and the lowest scored activities

  - The few differences which could not be resolved during discussions were reviewed and resolved by a knowledgeable senior manager

---

Technical Judgments
Use of the Questionnaire

A	We have *high confidence* that relevant time-dependent processes for the repository are captured in the measurement. Examples that would indicate high confidence include: (a) the PC measurement captures data from a closely related analogue system over time frames on the order of 10,000 years, (b) the PC measurement estimates the parameter changes by accelerating the time history, and that acceleration captures the relevant changes.
B	We have *moderate confidence* that relevant time-dependent processes for the repository are captured in the measurement. Examples that would indicate moderate confidence include: (a) the PC measurement captures data from loosely related analogue systems over time frames on the order of 10,000 years, (b) the PC measurement captures data from a closely related analogue system, but over time frames much greater than or much less than 10,000 years, (c) the PC measurement estimates the parameter changes by accelerating the time history, which causes the candidate parameter to change in a similarly representative manner to how it is expected to evolve in the repository environment.
C	We have *weak confidence* that relevant time-dependent processes for the repository are captured in the measurement. Examples that would indicate weak confidence include: (a) the PC measurement captures data from loosely related analogue system over time frames not representative of the 10,000 regulatory period, (b) the PC measurement estimates the parameter changes by accelerating time history, which causes the candidate parameter to change significantly differently than it is expected to evolve in the repository environment.
D	The PC measurement is designed to estimate post-closure changes through simple extrapolation of pre-closure measurement.

62

Slide Intentionally Blank

Performance Assessment Managers Provided the Necessary Management Value Judgments

Management value judgments

- Managers reviewed the overall process and endorsed the specific criteria being used to evaluate activities

- Managers answered a series of tradeoff questions, designed around the technical questions used in the questionnaire, to establish management value judgments about the relative importance of the criteria

- Management value judgment used in conjunction with the technical judgments to establish the overall utility for each activity

- Participants included the manager of the performance assessment project and the manager and/or deputy for related subprojects: natural systems, engineered systems, performance assessment strategy and scope, and the performance confirmation manager

Example Management Value Judgment for the Technical Judgment Question on Spatial Variability

(1 of 2)

- Participants reviewed the descriptions of the degree of confidence technical investigators may have that the measurements capture the spatial variability of the parameter - that is, the choices available for "technical judgment" of this question

3.2.a. Are the data from the PC activity representative of the spatial variability across the repository footprint, flow paths, or relevant spatial scale?

A	The data measures a parameter over all locations across the relevant spatial scale.
B	The data measures a parameter over representative locations we are *highly confident* represent the spatial variability across the relevant spatial scale.
C	The data measures a parameter over representative locations we are *moderately confident* represent the spatial variability across the relevant spatial scale.
D	The data measures a parameter over representative locations we are *weakly confident* represent the spatial variability across the relevant spatial scale.
E	The measurement give no information on the known spatial variability of the parameter across the relevant spatial scale and only measures a single (or non-representative few) location(s).

Example Management Value Judgment for the Technical Judgment Question on Spatial Variability

(2 of 2)

- Participants discussed the scale and assigned each of the five levels a weight indicative of relative accuracy of the measurement
- 8 participants
- Rankings highly consistent
- Average of the relative weights of the 8 participants used

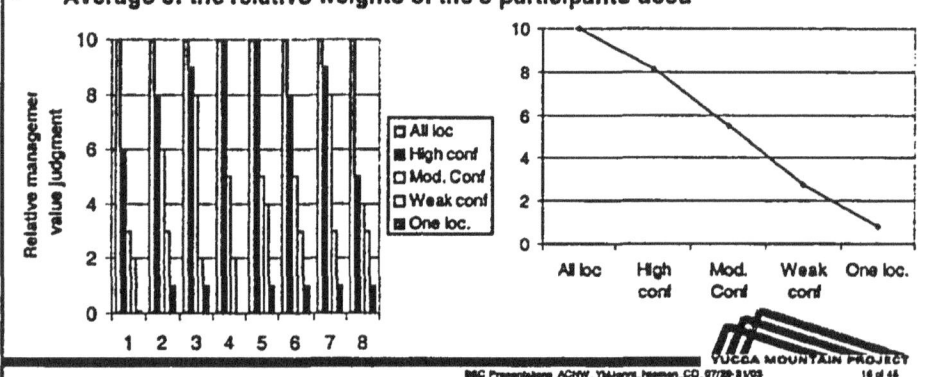

64

Example Management Value Judgment Accuracy

- "Value of perfect information" on a parameter was scaled by the estimated accuracy of the activity

- The three technical judgment aspects of accuracy were weighted by the management value judgments shown below:

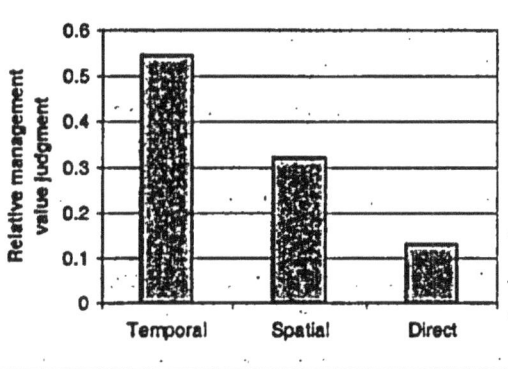

Management Value Judgments Related to Barrier Capability

- The contribution of "sensitivity to barrier capability" to total utility depends in part on the relative value assigned to each of the nine barriers

- Performance assessment managers assigned weights to each of the barriers, based on judgment:

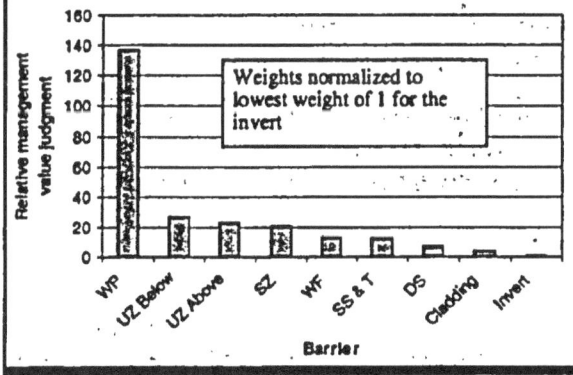

Weights normalized to lowest weight of 1 for the invert

- Informed by the risk prioritization report and the "one on" analyses

- Informed by discussions of barrier capability

Costs for Each Activity

- **Understanding both the benefits and the costs of a candidate activity is an essential component of the decision making process**

  - Including activities based solely on maximizing "benefit" may result in a highly cost-ineffective program

  - Including activities based solely on minimizing costs may leave highly valuable activities out

- **Costs are a consideration in developing portfolios, for example:**

  - Cost synergies may make combinations of activities more attractive

  - Costs can be a factor in deciding between otherwise equal activities

Phase 1 Summary

- **237 parameters and a total of 360 activities initially identified**

- **After discussion, evaluation, and consolidation, 204 parameters and 287 total activities remained**

- **A review meeting was held with representatives of the technical experts who provided input**

- **Technical experts indicated where they thought the results did not reflect their technical opinions, and comments were carried forward to the portfolio development phase**

"A Tale of Two Activities"
Phase 1, Activity Definition

- Activity 159a: Hydraulic testing of fault zone hydrologic characteristics, including anisotropy, in the saturated zone

- Technical judgments:
  - Saturated zone performance is highly sensitive to the parameter
  - Total system performance is very insensitive to the parameter
  - The conceptual model of the saturated zone flow is sensitive to changes in the parameter
  - Moderate to high confidence in the currently modeled range of the parameter
  - Parameter is not expected to vary temporally
  - High confidence that measurement captures the spatial variability in the parameter
  - Measurement is closely related to the parameter of interest

- Activity 28a: On-site testing of the hydrology, permeability, imbibition rate, and unsaturated hydraulic parameters of the invert materials

- Technical judgments:
  - Invert performance is moderately sensitive to the parameter
  - Total system performance is very insensitive to the parameter
  - The conceptual model of the invert flow is sensitive to changes in the parameter
  - Moderate to high confidence in the currently modeled range of the parameter
  - Parameter is expected to vary both during the pre- and the post-closure periods; measurements will not capture temporal changes
  - Low confidence that measurement captures the spatial variability in the parameter
  - Measurement is closely related to the parameter of interest

"A Tale of Two Activities"
Phase 1, Evaluation of Activities

A	We have *high confidence* that relevant time-dependent processes for the repository are captured in the measurement. Examples that would indicate high confidence include: (a) the PC measurement captures data from a closely related analogue system over time frames on the order of 10,000 years, (b) the PC measurement estimates the parameter changes by accelerating the time history, and that acceleration captures the relevant changes.
B	We have *moderate confidence* that relevant time-dependent processes for the repository are captured in the measurement. Examples that would indicate moderate confidence include: (a) the PC measurement captures data from loosely related analogue systems over time frames on the order of 10,000 years, (b) the PC measurement captures data from a closely related analogue system, but over time frames much greater than or much less than 10,000 years, (c) the PC measurement estimates the parameter changes by accelerating the time history, which causes the candidate parameter to change in a similarly representative manner to how it is expected to evolve in the repository environment.
C	We have *weak confidence* that relevant time-dependent processes for the repository are captured in the measurement. Examples that would indicate weak confidence include: (a) the PC measurement captures data from loosely related analogue system over time frames not representative of the 10,000 regulatory period, (b) the PC measurement estimates the parameter changes by accelerating time history, which causes the candidate parameter to change significantly differently than it is expected to evolve in the repository environment.
D	The PC measurement is designed to estimate post-closure changes through simple extrapolation of pre-closure measurement.

"A Tale of Two Activities"
Phase 1 – Operating Costs

Activity 159a

- Each test estimated to take 6 months to 1 year, total testing time 1 to 3 years

- Testing can be done using automated equipment in a shirtsleeve environment

- Estimated operating costs: $750,000

Activity 28a

- Testing estimated to take 6 months to 1 year

- Testing can be done using automated equipment in a shirtsleeve environment

- Estimated operating costs: $300,000

Approach

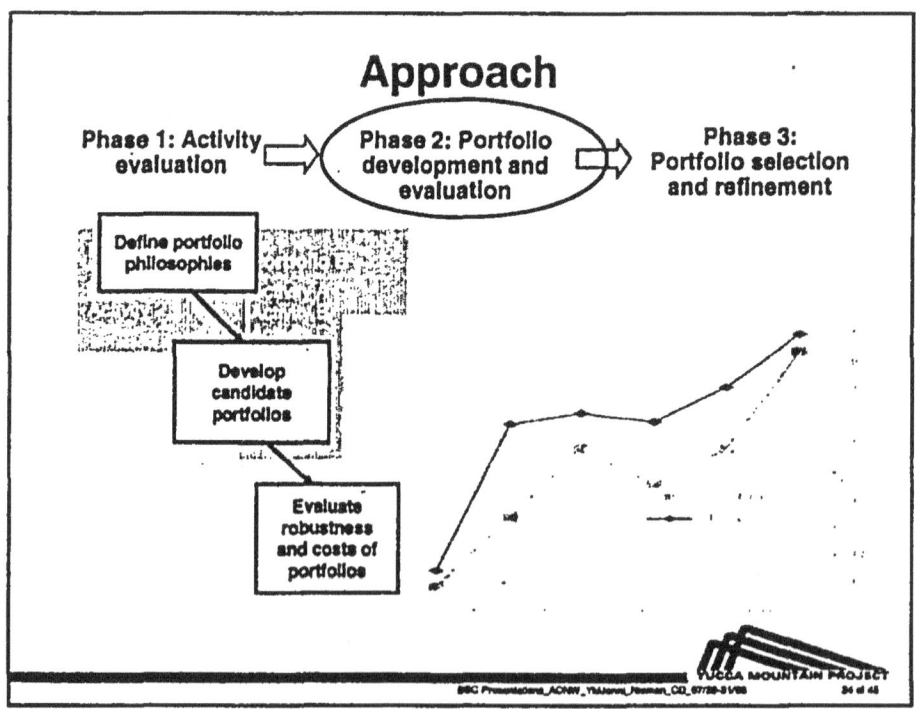

Phase 1: Activity evaluation

Phase 2: Portfolio development and evaluation

Phase 3: Portfolio selection and refinement

Define portfolio philosophies

Develop candidate portfolios

Evaluate robustness and costs of portfolios

Rationale for Portfolios

- Each candidate activity contributes to demonstrating compliance with one or more regulatory requirements

- The best portfolio does not necessarily result from ranking activities by utility, cost, or the ratio of utility to cost
  - Some regulatory requirements are not captured by the technical judgments and management value judgments input to the utility
  - Activity evaluations do not account for potential synergies

- Some costs cannot be assigned to individual activities (e.g., observation drift construction and remotely operated vehicle development)

- Portfolios of performance confirmation activities can be evaluated for regulatory compliance and for total cost

Philosophy for Portfolio Development

- Each portfolio addresses the performance confirmation requirements of 10 CFR 63

- Eleven portfolios were developed
  - Spanned a range of scope, costs, and robustness
  - Included portfolios that emphasized cost-benefit and hypothesis testing philosophies
  - Included portfolios that emphasized off-site work or on-site work

- Six of these portfolios were evaluated in detail
  - Scope, costs, robustness

Two Bounding Portfolios Were Developed

- All inclusive portfolio (K)
  - Includes all activities identified by the technical experts and evaluated as having positive benefit (ignoring costs)
- Minimum cost portfolio (A)
  - Least-cost set of activities that addresses the performance confirmation requirements of 10 CFR 63
  - The degree of activity for each 10 CFR 63 requirement is small, to achieve minimum cost
- These bounding portfolios were evaluated in detail
- A reduced version of the "all-inclusive" portfolio was developed, consisting of every parameter identified, but including only the most valuable activity associated with measuring that parameter (B)
  - This portfolio was not evaluated in detail

Cost Effectiveness Portfolios

- Three portfolios were developed
  - All activities were ranked by utility-to-cost ratio
  - "Threshold" utility-to-cost ratios were set for alternative portfolios (C, D, E)
  - Activities that met the threshold were included in the portfolio
  - Reviewed for cost synergies among activities
- Portfolios capturing 99 percent and 82 percent of the total potential utility were evaluated in detail

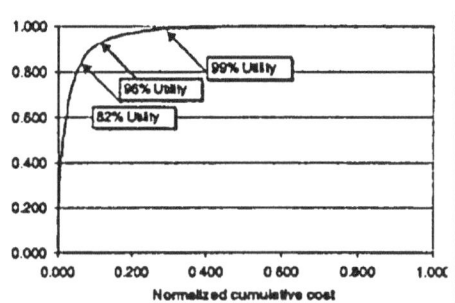

Hypothesis Testing Portfolios

- Two portfolios were defined around the notion of "hypothesis testing"
  - A set of performance "hypotheses" was developed at the barrier and total system level
  - Activities were identified as
    - Providing a direct test of an hypothesis
    - Providing an indirect test of an hypothesis (e.g., testing "inputs" to the hypothesis)
  - Example:
    - The surficial barrier will limit infiltration to less than nn percent of precipitation, averaged over the footprint and one year
- One hypothesis testing portfolio included only direct tests of the hypotheses (F)
- A second hypothesis testing portfolio included both direct and indirect tests of the hypotheses (G)
- Both portfolios were evaluated in detail

Type or Location Portfolios

- Three portfolios were developed that focus on either the type or the location of performance confirmation activities
  - Maximize use of a thermally accelerated emplacement drift (H)
    - Assumes a thermally accelerated drift will be included in the program; includes primarily activities making use of that drift
  - Maximize use of off-footprint testing (I)
    - Designed to keep worker risks as low as possible, and minimize interference of the program with activities in the Geologic Repository Operations Area
  - Maximize use of existing data, activities in existing facilities, and pre-emplacement activities (J)
    - Using data already collected or being collected in the Cross Drift Thermal Test and the Drift Scale Test
- These portfolios were not evaluated in detail
  - Did not provide significant additional benefit over other portfolios

Portfolio Evaluation Criteria

- Activities were mapped to the regulatory requirements in 10 CFR 63 Subpart F
  - Some activities support multiple requirements
- Attributes were totaled across the activities in each portfolio
  - Activity count
  - Total utility
  - Total operating plus capital cost
- Activity utilities were summed for each regulatory requirement in 10 CFR 63 Subpart F, within each portfolio
- A subjective assessment was made against each regulatory requirement in 10 CFR 63 Subpart F, for each portfolio
  - This added "coverage" as a subjective subcriterion

Six Portfolios Were Evaluated in Detail

- Minimum cost (Portfolio A)

- Cost effective - 82 percent total utility (Portfolio C)

- Cost effective - 99 percent total utility (Portfolio E)

- Hypothesis testing – Direct (Portfolio F)

- Hypothesis testing - Direct and indirect (Portfolio G)

- All inclusive (Portfolio K)

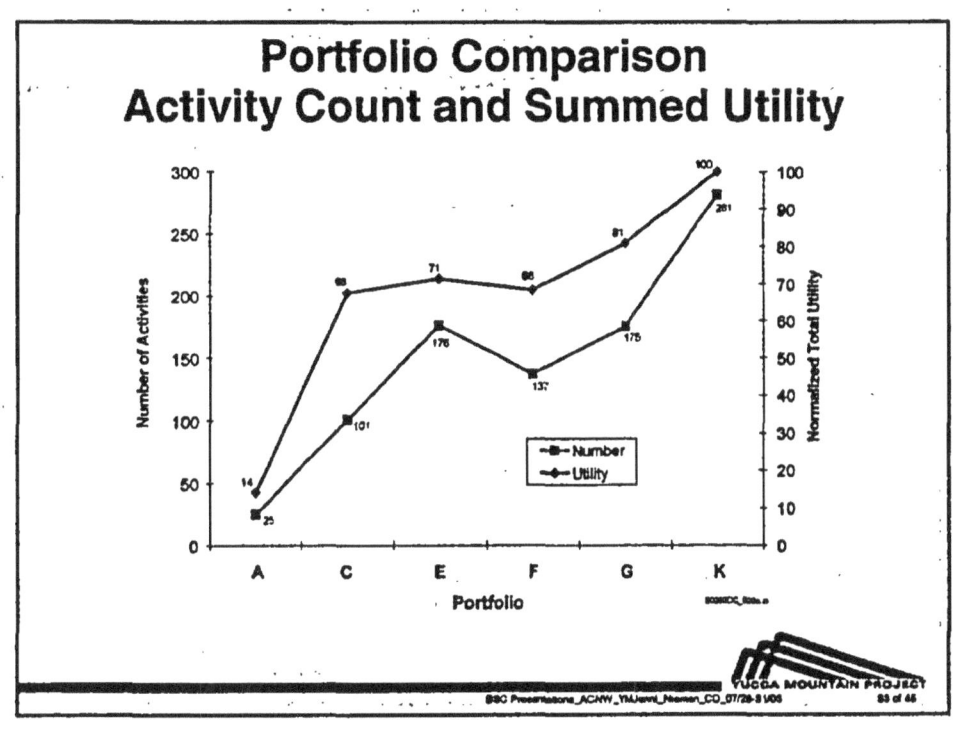

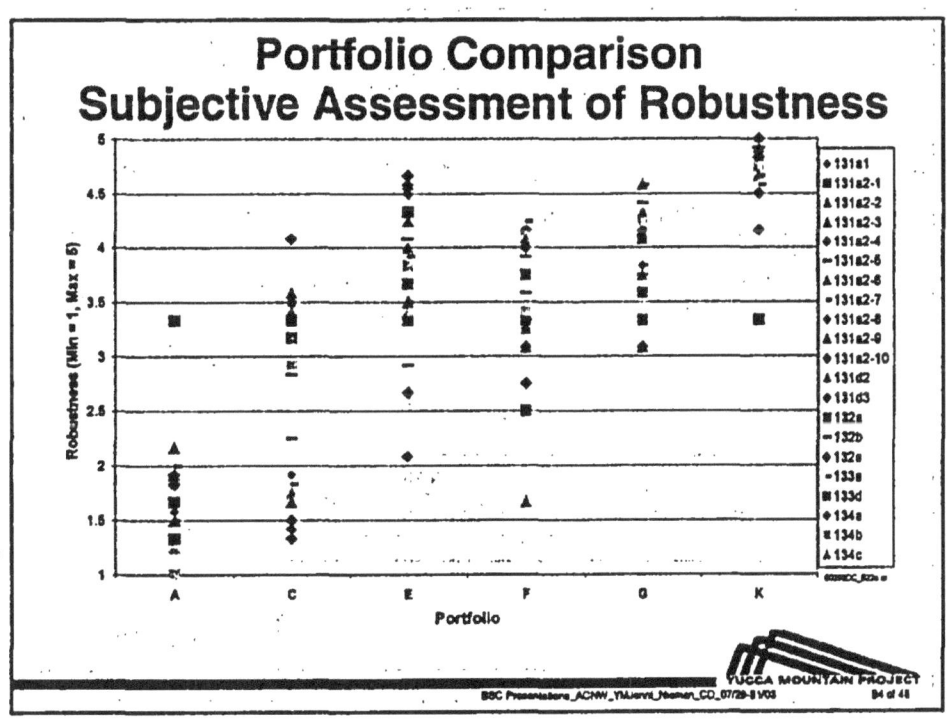

Portfolio Comparison
Relative Costs and Subjective Robustness

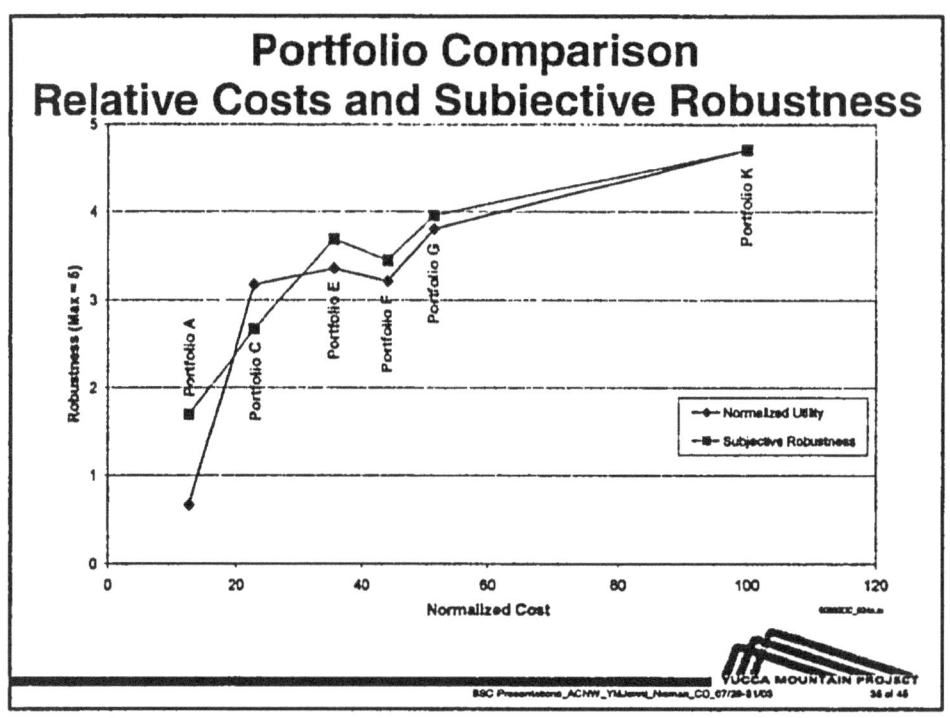

"A Tale of Two Activities"
Phase 2, Portfolio Development

- **Activity 159a Phase 1 Recap**
  - Hydraulic testing of fault zone hydrologic characteristics, including anisotropy, in the saturated zone
  - Total utility = 510
  - Estimated operating costs = $750,000

- **Activity 28a Phase 1 Recap**
  - On-site testing of the hydrology, permeability, imbibition rate, and unsaturated hydraulic parameters of the invert materials
  - Total utility = 1.7
  - Estimated operating costs = $300,000

- **The activities were included in the following portfolios:**

Activity	Portfolios										
	A	B	C	D	E	F	G	H	I	J	K
28a		X			X						X
159a		X	X	X	X		X		X		X

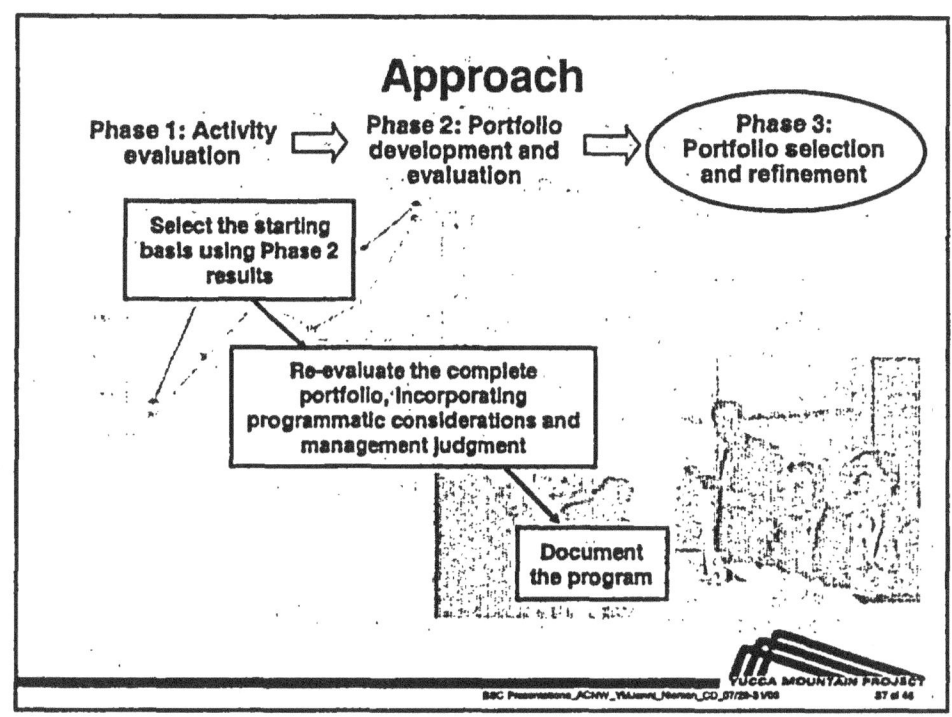

Approach

Phase 1: Activity evaluation ⇨ Phase 2: Portfolio development and evaluation ⇨ Phase 3: Portfolio selection and refinement

Select the starting basis using Phase 2 results

Re-evaluate the complete portfolio, incorporating programmatic considerations and management judgment

Document the program

Starting Basis

- **The BSC Manager of Projects and senior advisors**

  - Reviewed all eleven portfolios, and the detailed evaluation of six

  - Selected "Portfolio C" as the starting basis for the performance confirmation program

- **They directed several changes to that basis**

  - Activities were to be added to increase the robustness of the portfolio with respect to aspects of the regulation where it was judged relatively weaker than some other portfolios

  - Activities in the portfolio were described in terms of their relationship to the specific paragraphs of the regulatory requirement (10 CFR 63, Subpart F)

Portfolio Refinement

- **In a series of meetings, BSC senior management reviewed every activity in the modified basis portfolio, and made adjustments to the portfolio based on management judgment and programmatic considerations**

- **Of the initial 99 activities:**

  - **26 were removed from the portfolio because they were more logical candidates for other testing programs**

  - **3 were combined with other activities in the program based on the judgment that the combined activities were a more logical unit to consider**

  - **3 activities were retained in principle but modified in scope**

  - **2 new activities were added**

  * The Performance Confirmation Plan, Rev. 02 includes a description of the rationale for changes to the portfolio made during management discussions

---

"A Tale of Two Activities"
Phase 3, Portfolio Selection and Refinement

- **Phase 2 recap**

Activity	Portfolios											
	A	B	C	D	E	F	G	H	I	J	K	Performance confirmation program
28a: On-site testing of the hydrology, permeability, imbibition rate, and unsaturated hydraulic parameters of the invert materials	X			X						X		
159a: Hydraulic testing of fault zone hydrologic characteristics, including anisotropy, in the saturated zone		X	X	X	X		X		X		X	X (modified)

- **Portfolio C was selected as the starting basis for the performance confirmation program**

- **Adding Activity 28a would have increased the robustness with which one aspect of the regulation is met: confirming the performance of the invert barrier, but**

  - Portfolio C was already judged to be robust to that requirement

- **The scope of Activity 159a was increased during management discussions**

  - Expanded to include transport testing as well as flow testing

Backup

Backup: Modifications Made to Portfolio During Phase 3 (1 of 4)

#	Activities	Barrier	Rationale for Addition, Modification, or Removal
Modified Activities			
96b	Moisture content/potential in soil—in situ measurements with tensiometers, TDR and neutron probes, continuous monitoring	1	Modified: to be done only after significant rainfall events
159a	Fault zone hydrologic and transport characteristics (incl. anisotropy)—Fault hydraulic testing at 2 sites	4	Modified: expanded to include transport testing
185a	Number of waste packages hit in Zone 1—Modeling, analog studies	10	Modified: originally propose for Zones 1 and 2, reduced to apply to Zone 1 only
Added Items			
220a	Drift scale test in the lower lithophysal unit	2,3	Added to provide a test prior to construction authorization. Test not yet fully defined
221a	Geodetic monitoring of extensional tectonics in the Yucca Mountain Region	10	Added to provide additional indicator of igneous activity
Removed Items			
62a	Flow splitting and/or flow paths on all engineered barrier system surfaces—preemplacement test in drift with heat	5,6,9	More appropriate for the Scientific Testing and Evaluation Program
63a	Crack plugging—Laboratory Testing under controlled environment	5,6	More appropriate for the Scientific Testing and Evaluation Program
64a	Pit plugging—Laboratory Testing under controlled environment	5,6	More appropriate for the Scientific Testing and Evaluation Program
65b	Water flow rate through breaches in the engineered barrier system components—Laboratory test with heat	5,6	More appropriate for the Scientific Testing and Evaluation Program
78a	Flaws (including manufacturing flaws, and size, orientation, number)—Laboratory testing under controlled environment of specimens from manufacturing mockups and laboratory-prepared specimens	6	More appropriate for the Engineering Test and Evaluation Program

77

Backup: Modifications Made to Portfolio During Phase 3 (2 of 4)

#	Activities	Barrier	Rationale for Addition, Modification, or Removal
	Removed Items (continued)		
81b	Critical stress (K1SCC and stress threshold)—Laboratory testing under controlled environment of laboratory-prepared specimens and specimens from manufacturing mockups	6	More appropriate for either the Scientific Testing and Evaluation Program or the Engineering Test and Evaluation Program
95a	Physical/hydrological properties of soil—Core samples for measuring density, porosity and permeability	1	More appropriate for the Scientific Testing and Evaluation Program
98a	Matrix/fracture/bulk physical/hydro properties—Core samples for measuring density, porosity and permeability	1	More appropriate for the Scientific Testing and Evaluation Program
114b	Hydrologic and mineralogical properties of the PTn—Evaluation in alcoves from the shafts (Mapping, core samples, laboratory testing)	2	Appropriate as candidate for OCRWM's Science and Technology Program
135b	Hydrologic conditions beneath drift (drift shadow)—Analog studies, natural caves, old mines	3	Appropriate as candidate for OCRWM's Science and Technology Program
138a	Field Hydrologic properties of the CHn (and interface with TSw 3)	3	Appropriate as candidate for OCRWM's Science and Technology Program
139a	Hydrologic conditions CHn	3	Appropriate as candidate for OCRWM's Science and Technology Program
140a	Field sorptive characteristics of the CHn (including K_d)	3	Appropriate as candidate for OCRWM's Science and Technology Program
152a	K_d—Laboratory testing of rock matrix samples and alluvium samples	4	Appropriate as candidate for OCRWM's Science and Technology Program
154a	Recharge rates: regional model domain—Modeling and new field work (USGS regional model)	4	Appropriate as candidate for OCRWM's Science and Technology Program

Backup: Modifications Made to Portfolio During Phase 3 (3 of 4)

#	Activities	Barrier	Rationale for Addition, Modification, or Removal
	Removed Items (continued)		
156a	Flux at Site-Scale Model Boundaries—Use the coupled site/regional models to evaluate measured fluxes across boundaries—borehole dilution tests (concentration as a function of depth in the borehole, monitored over time)	4	Appropriate as candidate for OCRWM's Science and Technology Program
175b	EBS behavior under ground motion—Offsite shake table	5,6	More appropriate for either the Scientific Testing and Evaluation Program or the Engineering Test and Evaluation Program
176a	Alloy 22 failure criterion—Perform laboratory experiments on specimens of Alloy 22 with a range of residual stresses due to cold working/surficial damage	6	More appropriate for either the Scientific Testing and Evaluation Program or the Engineering Test and Evaluation Program
177a	Titanium grade 7 failure criterion—Perform laboratory experiments on specimens of Titanium grade 7 with a range of residual stresses due to cold working/surficial damage	5	More appropriate for either the Scientific Testing and Evaluation Program or the Engineering Test and Evaluation Program
183a	Dike system geometry—Analogs: mapping of exposed dike geometries, some drilling of dikes	10	Appropriate as candidate for OCRWM's Science and Technology Program
184a	Conduit system geometry—Field measurements, analog studies	10	Appropriate as candidate for OCRWM's Science and Technology Program
186a	Update modeling and laboratory experiments of damage to waste package from igneous event	6	Not needed – performance models treat waste package hit with magma as destroyed
188a	Ashplume: Incorporation ratio—Models and analogs, field studies	10	More appropriate for the Scientific Testing and Evaluation Program
189a	Ashplume: Waste particle size—Models and analogs	10	More appropriate for the Scientific Testing and Evaluation Program
195a	Proportion of eruptive styles—Models and analogs, field and laboratory measurements	10	Rolled into activity definition in 196a

Backup: Modifications Made to Portfolio During Phase 3 (4 of 4)

#	Activities	Barrier	Rationale for Addition, Modification, or Removal
	Removed Items (continued)		
196a	Distribution of magma type downdrift—Models and analogs	10	Appropriate as candidate for OCRWM's Science and Technology Program
197a	Distance magma travels downdrift—Models and analogs	10	Appropriate as candidate for OCRWM's Science and Technology Program
198a	Distribution of physical environment downdrift—Models and analogs	10	Appropriate as candidate for OCRWM's Science and Technology Program
213a	Dust Levels by Occupational Activity	10	Combined with activity 162a

YUCCA MOUNTAIN PROJECT

Left blank intentionally.

1.4 Elements of a Performance Confirmation Program - A Presentation of DOE's Selected Program and Its Components
James Blink, Bechtel SAIC Company; Lawrence Livermore National Laboratory

James Blink gave a talk about the elements of the Yucca Mountain performance confirmation program. He cautioned that some changes may occur during DOE's acceptance review and as the activities are developed for a license application. Dr. Blink noted that Phase 1 of the decision analysis was risk-based because it relied on performance assessment calculations. It was performance-based because it considered performance of the individual barriers and the total system. Phases 2 and 3 are considered risk-informed because they consider other factors such as relationships among activities, feasibility, operability, and cost.

The decision analysis focused the performance confirmation activities on the areas of highest risk. Three main groups (or classes) were identified:

- Disruptive scenario classes—igneous activity and seismic activity scenarios
- Biosphere-related activities—applicable to multiple scenarios
- Nominal scenario class (lower risk than the disruptive scenarios)—waste package and drip shield, pre-emplacement environment, land surface characteristics and the unsaturated zone below and above the proposed repository, coupled thermal processes, saturated zone, and cladding/waste form/invert

Igneous activity is the largest single contributor to the probability-weighted annual dose to the reasonably maximally exposed individual. The approximately 13 related performance confirmation activities will be designed to confirm the assumptions, data, and analyses of igneous events. Specific work may include:

- drilling of aeromagnetic anomalies to investigate possible buried volcanoes
- an updated expert elicitation on the probability of an igneous event
- further analyses of igneous consequences
- monitoring of regional extensional tectonics

Twenty-two activities have been proposed for performance confirmation of the waste package and drip shield. These will investigate mechanistic details of waste package and drip shield corrosion, and will involve lab tests on mockups to confirm stress sources as consequences of rockfalls and seismic activity. The near-field environments will be studied in thermally accelerated drifts using drift-end instruments, in-drift sampling, and a remotely operated vehicle. Parameters to measure will likely include temperature, humidity, dust and gas composition, pressure, radiolysis effects, condensate chemistry, thin film chemistry, and microbial activity. Radionuclides will be monitored in exhaust air to detect any waste package breach. The pressure seal of all waste packages can be measured with the remotely operated vehicle. This vehicle can also inspect emplacement drifts for ground support integrity and shape."

Other performance confirmation activities will relate to seismic activity (3), the biosphere (6), the pre-emplacement environment (8), coupled thermal processes (~12), and elements of unsaturated and saturated hydrology and chemistry (~13).

Left blank intentionally.

U.S. Department of Energy
Office of Civilian Radioactive Waste Management

Elements of the Yucca Mountain Performance Confirmation Program

Presented to:
Advisory Committee on Nuclear Waste

Presented by:
James A. Blink
Bechtel SAIC Company, LLC
Lawrence Livermore National Laboratory

July 29, 2003
Washington, D.C.

Purpose of This Presentation

- Describe the performance confirmation program proposed by BSC to DOE
    - Some changes may occur in the DOE acceptance process
    - Some evolution may occur as the activities are developed in preparation for the license application

Risk-Informed Perspective on the Performance Confirmation Program

- **Phase 1 of the decision analysis to scope the program was risk-based**
  - Relied on performance assessment calculations
- **Phase 1 of the decision analysis to scope the program was performance-based**
  - Considered performance of the individual barriers and the total system
- **Phases 2 and 3 of the decision analysis were risk-informed**
  - Included consideration of factors such as synergy among activities, feasibility, operability, and cost; in addition to the risk-based results of Phase 1
- **The resulting performance confirmation program is risk-informed, performance-based**

Risk-Informed Perspective on the Performance Confirmation Program
(Continued)

- **The performance confirmation program can be described from several viewpoints**
  - Time and location of implementation (Section 5, *Performance Confirmation Plan*, Rev 02)
  - Response to regulatory requirements of 10 CFR 63, Subpart F, and the *Yucca Mountain Review Plan* Section 2.4 (Section 4, *Performance Confirmation Plan*, Rev 02)
  - Association with repository barriers (Section 3 and Appendix B, *Performance Confirmation Plan*, Rev 02)
  - Risk-informed, performance-based terms, with respect to relationships to scenario classes, repository barriers, or processes
    - *This presentation is structured to reflect the risk-informed, performance-based program*
    - Risk is defined as the mean annual dose to the *reasonably maximally exposed individual*, calculated in total system performance assessment considering the probabilities of each scenario class

Organization of This Presentation

- The *Yucca Mountain Review Plan* Section 2.4.1 states the performance confirmation program should be "risk informed" and "*focused on parameters and natural and engineered barriers important to waste isolation*"

- The decision analysis focused the performance confirmation activities on the highest risk areas

- This presentation groups the activities into risk-informed categories
  - For convenience of discussion and to minimize repetition of activities
  - The groups are by total system performance assessment scenario class, barrier, and cross-cutting processes that affect a number of barriers

- The groups are sequenced with highest risk groups first and lowest risk groups last
  - Activities categorized in more than one group are described in detail in the group that best describes their primary performance confirmation role, and summarized in other groups

---

Activity Group Sequence

- Activities related to disruptive scenario classes (with highest risk scenario class first)
  - Igneous activity scenario class
  - Seismic activity scenario class

- Biosphere-related activities "downstream" of the nine barriers
  - These may apply to multiple scenario classes

- Nominal scenario class (which is lower risk than the disruptive scenario classes)
  - Waste package and drip shield
  - Preemplacement environment
  - Surface topography, soils, and bedrock; and the unsaturated zone (both above and below the repository)
  - Coupled thermal processes
  - Saturated zone
  - Cladding, waste form, and invert

Igneous Activity Scenario Class

- **Igneous activity is the largest single contributor to the probability-weighted annual dose to the reasonably maximally exposed individual**

- **Consequently, performance confirmation activities confirm assumptions, data, and analyses of igneous events**

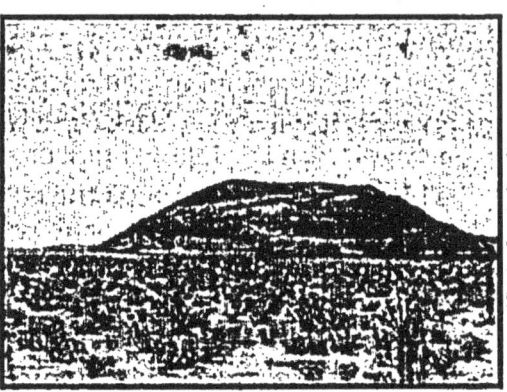

Igneous Activity Scenario Class
(Continued)

- **Probability of occurrence of igneous events**
  - Drilling of aeromagnetic anomalies (180a)
    - Improved data set
  - Updated expert elicitation (181a)
    - Incorporate improved data set

- **Consequences of igneous events**
  - Number of waste packages hit by magma (185a)
    - Calculations and analog studies
  - Behavior of contaminated ash (191a, 192a, 193a, 207a, 214a, 215a, 216a, 217a)
    - Ash loading, resuspension, redistribution, stabilization, and weathering
    - Radionuclide partition, sorption, dissolution/migration
    - Modeling, analogs, lab testing
  - Updated expert elicitation (182a)
    - Incorporate improved data set
- **Precursor conditions**
  - Satellite monitoring of regional extensional tectonics (221a)
    - Ongoing activity

Seismic Activity Scenario Class

- Seismic activity is expected to be a significant contributor to the probability-weighted annual dose to the reasonably maximally exposed individual

- Consequently, performance confirmation activities confirm assumptions, data, and analyses of seismic events

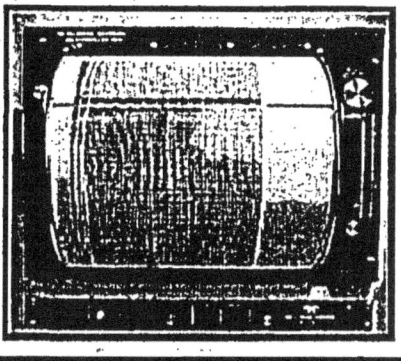

Seismic Activity Scenario Class
(Continued)

- Rock and soil dynamic properties at higher strains associated with major seismic events (173a)

  - Extend existing lower strain data set

- Regional seismic activity and near-field strong ground motions (167a)

  - Monitor for seismic activity and its consequences

  - Ongoing activity

- Inspection of surface and underground fault displacement in drifts if strong ground motion occurs (170a)

  - Contingency activity, using remotely operated vehicle

Biosphere-Related Activities "Downstream" of the Nine Barriers

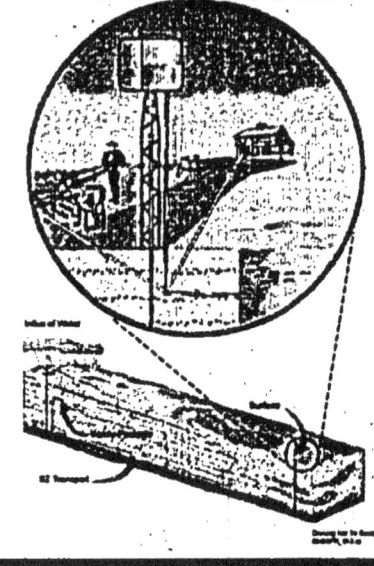

- Biosphere factors are potential multipliers on dose, without defense-in-depth mitigation

- During the long period of time prior to repository closure, human activities in the region are likely to change

- Consequently, performance confirmation activities confirm important biosphere factors

Biosphere-Related Activities "Downstream" of the Nine Barriers
(Continued)

- Periodic survey of *reasonably maximally exposed individual* characteristics and of occupational dust levels (162a)

  - Ongoing activity

- Natural analog study of the movement of radionuclides added to soil and their migration back to the water table, where they may be pumped back to the surface (166b)

  - Nominal and disruptive scenario classes

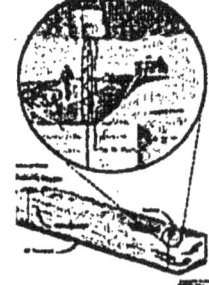

- Radionuclide movement to humans via plants (204a, 205a, 206a)

  - Nominal and disruptive scenario classes

- Radionuclide movement to humans through soil ingestion (direct or via animals) (208a)

  - Nominal and disruptive scenario classes

Waste Package and Drip Shield

- The waste package, in the environment created by the natural system, is expected to isolate radionuclides from the reasonably maximally exposed individual by preventing water from reaching the radionuclides

- The drip shield protects the waste package from rockfall and prevents advective transport from breached waste packages
  - Only the slower diffusive transport can operate under an intact drip shield

- Consequently, performance confirmation activities confirm assumptions, data, and analyses of waste package and drip shield performance

Waste Package and Drip Shield
Combined Activities

- Mechanistic details of waste package and drip shield corrosion (68a, 69a, 70a, 71a, 72a, 73a, 74a, 75a, 76a)
  - General corrosion, phase stability, localized corrosion, microbial corrosion
  - Ongoing activities
  - Strengthen extrapolation to 10,000 years
- Laboratory tests on mock-ups to confirm stress sources on the waste package and drip shield (79a)
  - Consequence of rockfall and seismic activity
- Waste package and drip shield environments (51a, 52a, 53a, 54e, 56e, 57a, 58e)
  - In thermally accelerated drifts, using drift-end instruments, in-drift samples, and the remotely operated vehicle
  - Includes temperature, humidity, dust composition, gas composition, pressure, radiolysis effects, condensate chemistry, thin film chemistry, and microbes
  - Temperature, humidity, and dust measurements include all emplacement drifts

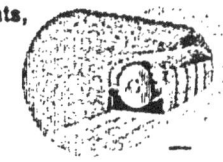

Waste Package

- **Monitoring radionuclides in exhaust air (251a)**
    - Measure at the end of each drift in a sensor module that also measures temperature and humidity
- **Pressure seal of all waste packages (83a)**
    - Measure with the remotely operated vehicle, imaging internal mechanical sensors that respond to equilibration of internal and external pressures

Both activities provide direct measures of overall waste package performance

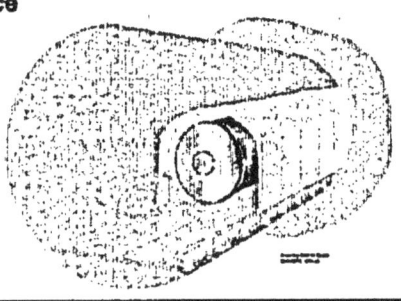

Drip Shield

- **Rockfall detection using acoustic/seismic tomography (59a1)**
    - Concept demonstrated by an existing university grant program
- **Inspection of drifts using the remotely operated vehicle (59a2)**
    - Drift 4 will include drip shields after about 5 years
    - Other drifts will be inspected for ground support integrity
- **Drift shape monitoring using the remotely operated vehicle in the thermally accelerated drifts (60b)**
    - Several concepts being considered

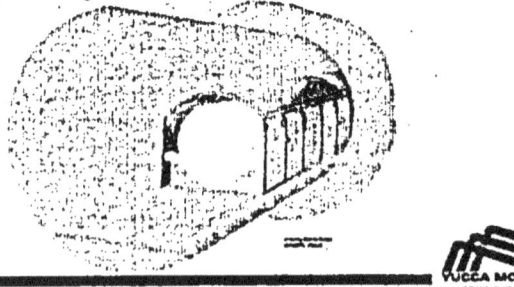

Preemplacement Environment

- The mechanical, hydrologic, and chemical environment in the emplacement drifts depends on the properties of the host rock in which the drifts are excavated

- Consequently, performance confirmation activities during construction of all emplacement drifts confirm host rock assumptions, data, and analyses

Preemplacement Environment
(Continued)

- Mapping of fractures, faults, stratigraphic contacts, and lithophysal characteristics (105a, 106a, 107a, 108a)

  - Three-pass construction

    - Excavate with light ground support
    - Remove Tunnel Boring Machine and map
    - Install permanent ground support

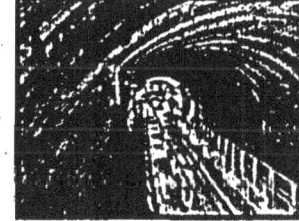

- Hydrologic proporties of significant fractures and faults (109a, 111b)

  - No characterization boreholes will be located over emplaced waste packages (gaps will be used, or characterization will use alcoves)

- Chemistry and age of pore water, using chloride mass balance and isotope chemistry (119a, 120a)

The Surface Barrier and the Unsaturated Zone Above and Below the Repository

- The surface topography, soils, and bedrock and the unsaturated zone above the repository limit the release of solubility-limited radionuclides (Pu and Np)
    - By reducing the rate and volume of water reaching the engineered barriers
    - By controlling the chemistry of water that reaches the engineered barriers
- The unsaturated zone below the repository reduces the annual dose in the event the drip shield and waste package barriers are breached (i.e., by an igneous event)
    - For short-lived radionuclides (such as Cs and Sr)
    - For solubility-limited radionuclides (such as Pu and Np)

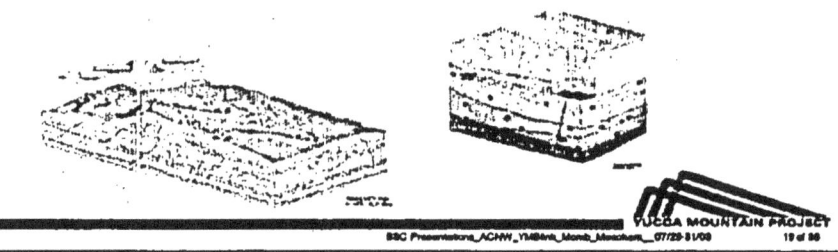

The Surface and the Unsaturated Zone Above the Repository

- Seepage into bulkheaded, low temperature alcoves (133b)
    - The situation most typical of the 10,000-year postclosure period
- Thermal seepage into an unventilated, thermally accelerated drift (51a, 133c1)
    - Detected by humidity change in the nearly stagnant, but slowly moving, air. Investigated using the remotely operated vehicle
    - Plausible because of the absence of ventilation, but unlikely due to elevated temperature
- Thermal seepage into ventilated heated drifts (51a, 133c2)
    - Detected by ventilation humidity change and investigated by the remotely operated vehicle
    - Unlikely due to ventilation and thermal effects
- Precipitation monitoring (84b)
    - To place seepage data in context
- Infiltration from rare high-intensity and long-duration storms (96b)
    - To place seepage data in context
- Seal performance (200a)
    - Seals prevent hydrologic short circuits

The Unsaturated Zone Below the Repository

- **Monitoring for radionuclides in deep boreholes near the footprint (151a)**
  - Confirms unsaturated zone barrier performance if engineered barriers fail
- **In situ test of transport and sorption properties of the unsaturated zone (137a)**
  - In a drift, prior to emplacement

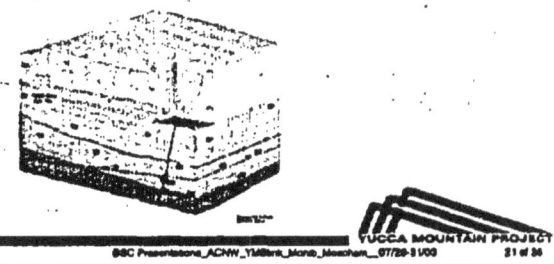

Coupled Thermal Processes

- **Heat added to the underground facilities by radionuclide decay will elevate temperatures for long periods**
  - Elevated temperatures drive thermal-hydrologic-mechanical-chemical processes in the drift and near-field rock
- **Consequently, performance confirmation activities confirm the assumptions, data, and analyses of coupled thermal processes**

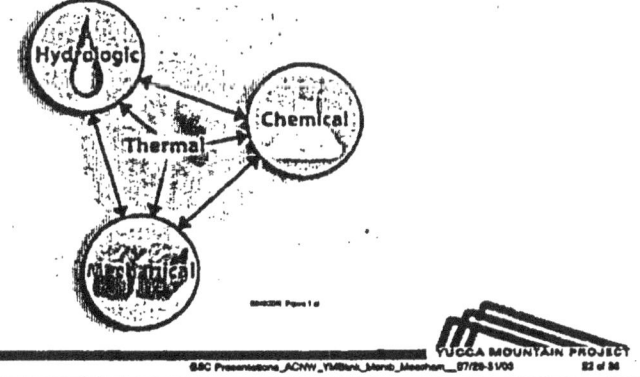

Coupled Thermal Processes
(Continued)

- Lower lithophysal drift scale test prior to emplacement (220a)
  - In the cross drift that was excavated by a tunnel boring machine
  - Thermal and thermal-mechanical processes are primary objectives; thermal-hydrologic and thermal-chemical processes are secondary objectives
- Drift 3, thermally accelerated by ventilation control (125a, 128a, 129b, 131a)
  - Near-field focus, uses an observation drift rather than in-drift boreholes
  - Fracture permeability, rock saturation, temperature, and water chemistry
- Drift 4, thermally accelerated by waste package aging and derating (51a, 52a, 53a, 54e, 56e, 57a, 58e)
  - Engineered barrier environment focus using the remotely operated vehicle
  - Includes drip shields and termination of ventilation at 5 years

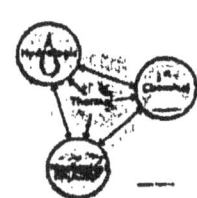

Saturated Zone

- The saturated zone reduces the annual dose in the event the drip shield and waste package barriers are breached (i.e., by an igneous event)
  - For short-lived radionuclides (such as Cs and Sr)
  - For solubility-limited radionuclides (such as Pu and Np)
- Consequently, performance confirmation activities confirm the assumptions, data, and analyses of the saturated zone

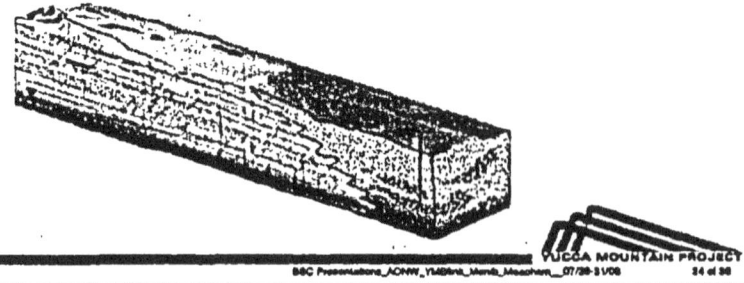

94

Saturated Zone
(Continued)

- Monitoring for radionuclides in deep boreholes downstream from the footprint (151a)
  - Confirms unsaturated and saturated zone barrier performance if engineered barriers fail
- Saturated zone chemistry and water levels (150a)
  - Chemistry affects retardation
  - Water levels are diagnostic of flow paths and rates
- Saturated zone colloids (153a)
  - Laboratory studies using field samples
- Saturated zone fault zone hydrology (159a)
  - Deep borehole tests
  - Faults affect flow paths and rates

Cladding, Waste Form, and Invert

- The cladding, waste form, and invert are barriers important to waste isolation, and contribute to defense-in-depth, but they are less important to annual dose than other barriers and processes

- Consequently, less emphasis is placed on confirmation of these barriers

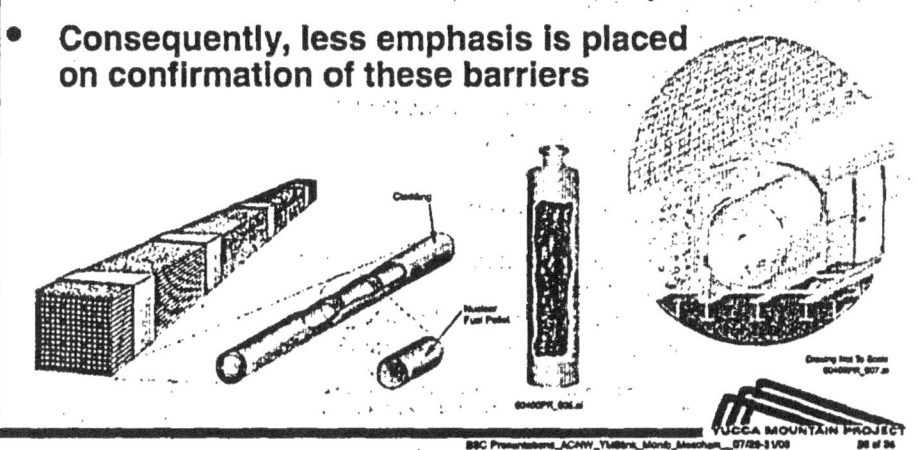

Cladding, Waste Form, and Invert
(Continued)

- **Radionuclide inventory (199a)**
  - From waste acceptance documents
- **Sorption coefficients for waste form colloids (16a)**
  - Laboratory tests
- **Monitor cladding studies (1a)**
  - From dry storage facilities
  - From academic and industrial research
- **Measure invert tuff gravel sorption coefficients (36a)**
  - Laboratory tests

The Performance Confirmation Program Focuses on Importance to Waste Isolation

Number of Activities

- Igneous activity scenario class (13 activities)
- Seismic activity scenario class (3)
- Biosphere-related activities (6)
- Waste package and drip shield (22)
- Preemplacement environment (8)
- Surface barrier and the unsaturated zone (8+1*)
- Coupled thermal processes (5+7*)
- Saturated zone (3+1*)
- Cladding, waste form, and invert (4)

Scenario classes that contribute most to risk are well represented in the performance confirmation program

Barriers that contribute most to risk are well represented

Barriers that contribute least to risk are represented minimally

Caveat: The 72 activities have varying degrees of scope complexity and cost
* The second number indicates activities included in a prior group

Backup

Performance Confirmation Activities - 1 of 4

- 1a—Monitoring the literature regarding commercial spent nuclear fuel cladding during the preclosure period, including tracking empirical data on cladding failure in dry storage facilities as well as academic and industrial research on mechanistic processes affecting cladding degradation
- 16a—Laboratory testing of sorption coefficients (K_ds) for waste form colloids
- 36a—Laboratory testing of invert chemistry and sorption coefficients (K_ds)
- 51a—Monitoring of the air temperature and relative humidity at the exit of all emplacement drifts
- 52a—Monitoring and laboratory testing of quantity and composition of dust on engineered barrier surfaces in a thermally accelerated emplacement drift
- 53a—Monitoring and laboratory testing of the quantity and composition of dust in the air in the emplacement drifts
- 54e—Monitoring of gas composition, pressure, and radiolysis effects within a thermally accelerated emplacement drift using a remotely operated vehicle
- 56e—Monitoring, sampling, and laboratory testing of condensation water quantities, composition, and ionic characteristics, including microbial effects, from a thermally accelerated emplacement drift
- 57a—Laboratory testing of water conditions, including thin films, on engineered barrier system components
- 58e—Monitoring, sampling, and laboratory testing of microbial types and amounts on engineered barrier surfaces in a thermally accelerated emplacement drift
- 59a1—Rockfall monitoring and aboveground motion sensing throughout the underground facility using acoustic or seismic tomography with sensors located in accessible areas, which can also measure strong ground motion
- 59a2—Inspection of the underground facility, waste package and other engineered components, with a remotely operated vehicle, when indicated by the results of the acoustic or seismic monitoring of the underground facility
- 60b—Monitoring drift shape, drift degradation, waste package, and drift components of a thermally accelerated emplacement drift with a remotely operated vehicle
- 68a—Laboratory testing of passive current density on Alloy 22 and Titanium Grade 7
- 69a—Laboratory testing of the weight loss rate of Alloy 22 and Titanium Grade 7
- 70a—Laboratory testing of surface dissolution of Alloy 22 and Titanium Grade 7
- 71a—Laboratory testing of surface composition and passive film of Alloy 22 and Titanium Grade 7 coupons from a thermally accelerated emplacement drift

Performance Confirmation Activities - 2 of 4

- 72a—Laboratory testing of the mechanical properties of passive film on Alloy 22 and Titanium Grade 7 coupons from a thermally accelerated emplacement drift
- 73a—Laboratory testing and analysis of phase transformations of Alloy 22 coupons from a thermally accelerated emplacement drift
- 74a—Laboratory testing and analysis of the open circuit potential of Alloy 22 and Titanium Grade 7
- 75a—Laboratory testing and analysis of the critical potential of Alloy 22 and Titanium Grade 7
- 76a—Laboratory testing and analysis of the critical ionic concentration, both abiotic and biotic, on Alloy 22 and Titanium Grade 7
- 79a—Laboratory analysis of waste package and drip shield stress sources using Alloy 22 and Titanium Grade 7 specimens and manufacturing mockups
- 83a—Monitoring the internal pressure of the waste packages using mobile radiation detectors to detect the shadow of pressure-sensitive internal sensors
- 84b—Precipitation monitoring and analysis of precipitation composition
- 96b—Measurements of moisture content and potential in surface soils after significant rainfall events
- 105a—Mapping of fracture characteristics in all drifts and shafts during repository construction
- 106a—Mapping of fault zone characteristics in all drifts and shafts during repository construction
- 107a—Mapping of stratigraphic contacts of geologic units in all drifts and shafts during repository construction, including revisiting the geologic framework model if necessary
- 108a—Mapping of lithophysal characteristics in all drifts and shaft walls within the lithophysal host rock units during repository construction
- 109a—Evaluation of the hydrologic properties of fractures using a combination of gas and liquid tracer tests as well as laboratory testing of moisture retention properties of the fractures
- 111b—Evaluation of the hydrologic properties of any previously undetected faults found during repository construction
- 119a—Laboratory analysis of chloride mass balance, based on samples taken throughout the underground facility
- 120a—Laboratory analysis of isotope chemistry (U, Sr, O, H, ^{36}Cl, ^{3}H, C) within the unsaturated zone, based on samples taken throughout the underground facility
- 125a—Monitoring of rock mass moisture content in boreholes in the near-field rock of a thermally accelerated emplacement drift

Performance Confirmation Activities - 3 of 4

- 128a—Air permeability testing to measure fracture permeability in the near-field rock of a thermally accelerated emplacement drift
- 129b—Monitoring of temperatures and thermal gradients in the near-field rock of a thermally accelerated emplacement drift
- 131a—Collection and laboratory analysis of water chemistry in the near-field rock of a thermally accelerated emplacement drift
- 133b—Monitoring, collection, and laboratory analysis of seepage water from bulkheaded alcoves on the intake side of the repository
- 133c1—Monitoring, collection, and laboratory analysis of seepage water from a thermally accelerated drift, using a remotely operated vehicle
- 133c2—Monitoring, collection, and laboratory analysis of seepage water from emplacement drifts, using a remotely operated vehicle
- 137a—Testing of transport properties and field sorptive properties of the crystal-poor member of the Topopah Spring Tuff (Tptp)
- 150a—Monitoring, sampling, and analyzing saturated zone water from Nye County and site wells for water levels, Eh, and pH
- 151a—Monitoring, sampling, and analyzing saturated zone water from Nye County and site wells for radionuclide concentrations
- 153a—Laboratory studies of the characteristics of natural colloids from saturated zone water samples, including colloid concentrations, particle size distribution, and mineralogy
- 159a—Hydraulic testing of fault zone hydrologic characteristics, including anisotropy, in the saturated zone
- 162a—Periodic surveys of the habitats and characteristics of the reasonably maximally exposed individual and dust levels associated with occupational activity
- 166b—Natural analogue studies of the fraction of radionuclides from the soil captured by the water table
- 167a—Monitoring regional seismic activity, if such data are not available through other programs
- 170a—Observation of subsurface and surface fault displacement after significant local or regional seismic events
- 173a—Laboratory testing of rock and soil dynamic properties using higher strains than have been tested during site characterization
- 180a—Drilling of aeromagnetic anomalies for volcanic event count modeling

Performance Confirmation Activities - 4 of 4

- 181a—Update probability estimates for volcanic intrusion by updating the probabilistic volcanic hazard analysis using expert elicitation
- 182a—Update estimated consequences of an igneous intrusion using expert elicitation
- 185a—Updated modeling and analogue studies of the number of waste packages hit from igneous events
- 191a—Updated modeling and analogue studies of initial mass loading of ash
- 192a—Field measurements of the resuspension and redistribution of volcanic ash in analogues
- 193a—Experimental and analogue studies of the resuspension and redistribution of ash resulting from human activities (e.g., plowing)
- 199a—Monitoring of average codisposal and commercial spent nuclear fuel waste package radionuclide inventory by tracking the waste stream receipt certification
- 200a—Laboratory testing of effectiveness of ramp, borehole, and shaft seals prior to submitting a license amendment to receive and possess waste
- 204a—Laboratory testing and literature review of radionuclide transfer factors, root uptake
- 205a—Laboratory testing and literature review of radionuclide foliar translocation factor
- 206a—Laboratory testing and literature review of radionuclide foliar interception factor
- 207a—Laboratory testing of sorption coefficients (K_ds) for ash particles in soils
- 208a—Laboratory testing for inadvertent soil intake containing radionuclides by humans and animals
- 214a—Laboratory testing for radionuclide activity partition by ash and soil particle size
- 215a—Laboratory testing and literature review of airborne volcanic ash level stabilization
- 216a—Laboratory testing for waste particle dissolution and migration in ash and soil
- 217a—Analysis of ash particles for dimensional changes due to weathering
- 220a—Drift Scale Test in the lower lithophysal unit
- 221a—Geodetic monitoring of extensional tectonics in the Yucca Mountain region using global positioning system satellite monitoring as a potential indicator of future igneous activity
- 251a—Monitoring of ventilation system exhaust gas for radionuclides

Performance Confirmation Activities and Regulatory Requirements - 1 of 5

- 10 CFR 63.131(a)(1)
  - "The performance confirmation program must provide data that indicate, where practicable, whether: Actual subsurface conditions encountered and changes in those conditions during construction and waste emplacement operations are within the limits assumed in the licensing review"
  - 51a, 52a, 53a, 54e, 56e, 58e, 59a1, 59a2, 60b, 105a, 106a, 107a, 108a, 109a, 111b, 119a, 120a, 125a, 128a, 129b, 131a, 133b, 133c1, 133c2
- 10 CFR 63.131(a)(2)—Total system performance, nominal scenario class
  - Directly affects total system performance, not through a barrier: "The performance confirmation program must provide data that indicate, where practicable, whether: ...Natural and engineered systems and components required for repository operation, and that are...assumed to operate as barriers after permanent closure, are functioning as intended and anticipated"
  - 83a, 151a, 251a
- 10 CFR 63.131(a)(2)—Surface topography, soils and bedrock barrier
  - 51a, 84b, 96b, 133b, 133c1, 133c2
- 10 CFR 63.131(a)(2)—Unsaturated zone above the repository barrier
  - 51a, 105a, 106a, 107a, 108a, 109a, 111b, 119a, 120a, 125a, 128a, 129b, 131a, 133b, 133c1, 133c2, 220a
- 10 CFR 63.131(a)(2)—Unsaturated zone below the repository barrier
  - 105a, 106a, 107a, 108a, 109a, 111b, 119a, 120a, 125a, 128a, 131a, 137a, 151a, 220a

- 10 CFR 63.131(a)(2)—Saturated zone between the repository and the accessible environment barrier
  - 150a, 151a, 153a, 159a
- 10 CFR 63.131(a)(2)—Drip shield barrier
  - 53a, 54e, 56e, 57a, 59a1, 59a2, 60b, 68a, 69a, 70a, 74a, 75a, 76a, 79a
- 10 CFR 63.131(a)(2)—Waste package barrier
  - 51a, 52a, 53a, 54e, 56e, 57a, 58e, 59a1, 59a2, 68a, 69a, 70a, 71a, 72a, 73a, 74a, 75a, 76a, 79a, 129b, 133b, 133c1, 133c2
- 10 CFR 63.131(a)(2)—Commercial spent nuclear fuel cladding barrier
  - 1a
- 10 CFR 63.131(a)(2)—Waste form barrier
  - 16a, 199a
- 10 CFR 63.131(a)(2)—Drift invert barrier
  - 36a
- 10 CFR 63.131(a)(2)—Total system performance, disruptive scenario classes
  - Directly affects system performance, not through a barrier
  - 162a, 166b, 167a, 170a, 173a, 180a, 181a, 182a, 185a, 191a, 192a, 193a, 204a, 205a, 206a, 207a, 208a, 214a, 215a, 216a, 217a, 221a

---

- 10 CFR 63.131(d)(2)
  - "The program must be implemented so that: It provides baseline information and analysis of that information on those parameters and natural processes pertaining to the geologic setting that may be changed by site characterization, construction, and operational activities"
  - 51a, 52a, 53a, 54e, 56e, 58e, 59a1, 59a2, 60b, 96b, 105a, 106a, 107a, 108a, 109a, 111b, 119a, 120a, 125a, 128a, 129b, 131a, 133b, 133c1, 133c2, 150a, 151a
- 10 CFR 63.131(d)(3)
  - "The program must be implemented so that: It monitors and analyzes changes from the baseline condition of parameters that could affect the performance of a geologic repository"
  - 51a, 52a, 53a, 54e, 56e, 58e, 59a1, 59a2, 60b, 84b, 96b, 105a, 106a, 107a, 108a, 109a, 111b, 119a, 120a, 125a, 128a, 129b, 131a, 133b, 133c1, 133c2, 150a, 151a, 167a, 170a
- 10 CFR 63.132(a)
  - "During repository construction and operation, a continuing program of surveillance, measurement, testing, and geologic mapping must be conducted to ensure that geotechnical and design parameters are confirmed and to ensure that appropriate action is taken..."
  - 51a, 52a, 53a, 54e, 56e, 58e, 59a1, 59a2, 60b, 105a, 106a, 107a, 108a, 125a, 128a, 129b, 131a, 133b, 133c1, 133c2, 167a, 170a, 173a
- 10 CFR 63.132(b)
  - "Subsurface conditions must be monitored and evaluated against design assumptions"
  - 51a, 52a, 53a, 54e, 56e, 58e, 59a1, 59a2, 60b, 125a, 129b, 131a, 133b, 133c1, 133c2

Performance Confirmation Activities and Regulatory Requirements - 4 of 5

- 10 CFR 63.132(e)
  - "In situ monitoring of the thermomechanical response of the underground facility must be conducted until permanent closure, to ensure that the performance of the geologic and engineering features is within design limits"
  - 51a, 59a1, 59a2, 60b, 129b, 220a

- 10 CFR 63.133(a)
  - "During the early or developmental stages of construction, a program for testing of engineered systems and components used in the design, such as, for example, borehole and shaft seals, backfill, and drip shields, as well as the thermal interaction effects of the waste packages, backfill, drip shields, rock, and unsaturated zone and saturated zone, must be conducted"
  - 1a, 16a, 36a, 51a, 52a, 53a, 54e, 56e, 57a, 58e, 59a1, 59a2, 60b, 68a, 69a, 70a, 71a, 72a, 73a, 74a, 75a, 76a, 79a, 125a, 128a, 129b, 131a, 133c1, 133c2, 167a, 170a, 199a, 200a, 220a

- 10 CFR 63.133(d)
  - "Tests must be conducted to evaluate the effectiveness of borehole, shaft, and ramp seals before full-scale operation proceeds to seal boreholes, shafts, and ramps"
  - 200a

- 10 CFR 63.134(a)
  - "A program must be established at the geologic repository operations area for monitoring the condition of the waste packages. Waste packages chosen for the program must be representative of those to be emplaced in the underground facility"
  - 83a, 151a, 251a

Performance Confirmation Activities and Regulatory Requirements - 5 of 5

- 10 CFR 63.134(b)
  - "Consistent with safe operation at the geologic repository operations area, the environment of the waste packages [chosen for the program] must be representative of the environment in which wastes are to be emplaced"
  - 51a, 52a, 53a, 54e, 56e, 57a, 58e, 59a1, 59a2, 133b, 133c1, 133c2

- 10 CFR 63.134(c)
  - "The waste package monitoring program must include laboratory experiments that focus on the internal condition of the waste packages. To the extent practical, the environment experienced by the emplaced waste...must be duplicated in the laboratory experiments"
  - 1a, 16a, 69a, 71a, 72a, 73a

Left blank intentionally.

4.5 Documentation and Further Development of the Performance Confirmation Program
A Presentation on Possible Changes in the Next Revision of DOE's Performance Confirmation Plan
Deborah Barr, U.S. Department of Energy

Deborah Barr described how the PC program will likely evolve. Revision 3 of the Performance Confirmation Plan is scheduled for the spring of 2004. This revision will include specific details about the plans, including:

- specific activities (what, when, where, and how)
- baseline established for PC
- bounds and tolerances for parameters
- management and administration of the PC program
- test plans
- the process for reporting variances and appropriate corrective actions

Some of the proposed PC activities will require feasibility evaluation or even the development of new technology.

Left blank intentionally.

Documentation and Further Development of the Performance Confirmation Program

Presented to:
Advisory Committee on Nuclear Waste

Presented by:
Deborah Barr
Office of Repository Development
Office of License Application and Strategy
U.S. Department of Energy

July 29, 2003
Washington D.C.

Path Forward - Revision 2

- Revision 2 of the *Performance Confirmation Plan* is currently in U.S. Department of Energy review
    - U.S. Department of Energy review completion - August 2003
    - Changes and corrections (if necessary) - September 2003

Path Forward - Revision 3

- **Revision 3 of the *Performance Confirmation Plan* is scheduled for spring of 2004**
  - Define activities (what, when, where, and how)
  - Establish expected baseline for performance confirmation activities
  - Establish bounds and tolerances for parameters
  - Management and administration
  - Identify needed test plans
  - Define process for reporting variances and describe the appropriate corrective action steps

* The following slides will give more details on each of the above bullets

Path Forward - Revision 3
(Continued)

- **Define activities (what, when, where, and how)**
  - Crosswalk to current and previous testing
  - Specify the spatial range over which data will be collected
  - Specify whether data needs to be collected continuously or at specified time intervals
  - Specify whether data will be collected using a remotely operated vehicle, in a laboratory setting, or with persons wearing personal protective equipment
  - Specify the type of power and communication instrumentation needed

Path Forward - Revision 3
(Continued)

- **Establish expected baseline for performance confirmation activities**
- **Establish bounds and tolerances for parameters**

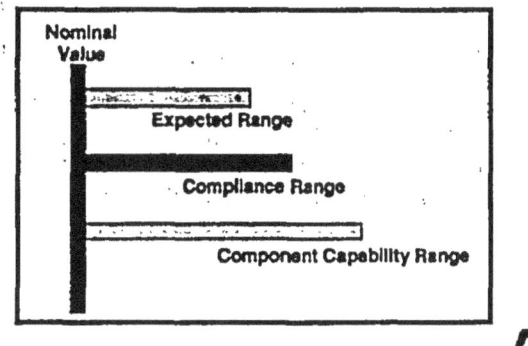

Path Forward - Revision 3
(Continued)

- **Management and administration**
  - **Identify general test procedures**
  - **Organizational structures for conducting the program**
- **Identify needed test plans ("one-time" tests and multiple tests)**
  - **Adequate level of detail on activity definitions to implement tests**
  - **Establish test decommissioning process**

Path Forward - Revision 3
(Continued)

- **Define process for reporting variances and describe the appropriate corrective action steps**

  - Routine reporting (all tests)

  - Variance analysis based on data trends and forecasts

  - Reporting of actual data outside regulatory limits

  - Corrective actions can include model improvements, test modifications, repository design/construction changes, removal of waste packages, waste retrieval (all in conjunction with NRC and stakeholder reporting and interaction)

Path Forward - Revision 3
(Continued)

- **Provide design requirements and further details on:**

  - Accelerated drift tests

    - Drift scale test in the lower lithophysal unit

    - Thermally accelerated drift focused on near-field coupled processes

    - Thermally accelerated drift focused on in-drift coupled processes

  - Exhaust mains instrumentation/monitoring systems

  - Seepage/H_2O collection system

  - Rockfall monitoring system

Path Forward - Implementation

- Implement *Performance Confirmation Plan*
  - Monitor, test, and collect data
  - Analyze and evaluate data
  - Take corrective actions should significant variances arise

Technology Development Areas

- Several performance confirmation activities require feasibility evaluation and/or technology adaptation/development
  - Remotely operated vehicle (with reduced dependence on infrastructure)
  - Radionuclide sensors with increased sensitivity (e.g., measuring in the exhaust mains)
  - Seepage detection via humidity spikes
  - Rockfall or engineered barrier system collapse detection via acoustic/ seismic tomography
  - Waste package hermetic seal via non electronic internal pressure sensors
  - Fast, effective mapping
  - Automated monitoring of drift deformation
- The performance confirmation staff is currently pursuing each of these areas
  - Some activities may be deleted and replaced as a result

Upcoming Milestones

- *Performance Confirmation Plan* Rev 03 - March 2004
- *Safety Analysis Report,* Chapter 4 - December 2004

.6 NRC's Risk Insights Initiative and Its Impact on Review of Performance Confirmation Plans - Risk-Informing Performance Confirmation
Timothy McCartin, U.S. Nuclear Regulatory Commission

imothy McCartin gave a talk about the process of risk-informing performance confirmation.)OE is required to identify and describe repository barriers. Under a risk-informed approach,)OE would identify the relative risk significance of each barrier. Mr. McCartin described the pproach as an iterative one in which risk significance is described, a quantitative basis is rovided, uncertainties are considered, important parameters and assumptions are identified, nd confirmatory evidence is considered.

Ir. McCartin gave a hypothetical example indicating that retardation in alluvial deposits is risk ignificant. Alluvium has the potential to delay the movement of most radionuclides for very)ng time periods. There is little uncertainty about the retardation of I-129 or Tc-99. These idionuclides are highly mobile and move with the water. Am-241 and Pu-240 tend to be elatively immobile under most circumstances. However, the retardation factor for Np-237 is ighly variable such that this radionuclide could be retarded for a few centuries or more than 00,000 years.

Ir. McCartin noted that risk insights identify areas to be considered for performance onfirmation. An NRC risk insights report is now being prepared that is based on a risk aseline, provides quantitative bases for relative risk, and identifies further calculations that iay be needed. The risk insights report will be updated in the future as needed.

Left blank intentionally.

Risk-Informing
Performance Confirmation

144th Meeting of
Advisory Committee on Nuclear Waste
(Working Group on Performance Confirmation Plans)
July 30, 2003

Tim McCartin 301-415-7285 tjm3@nrc.gov
Dave Esh 301-415-6705 dwe@nrc.gov
Division of Waste Management
U.S. Nuclear Regulatory Commission

Outline

- Performance Confirmation Perspective
- Approach
- Engineered Barrier Example
- Natural System Example
- Summary

08/13/2003

Performance Confirmation

◆ Evaluate adequacy of information used to demonstrate compliance
  - subsurface conditions are within the limits assumed during licensing review
  - barriers functioning as intended and anticipated
◆ Provide data where practicable
◆ In situ monitoring, laboratory and field testing, and in situ experiments

08/13/2003

Risk Informed

◆ Risk significance of each barrier

◆ Uncertainty in estimating performance of barriers

Note:
DOE required to identify and describe repository barriers

08/13/2003

Overall Approach
(Iterative)

Describe Risk Significance

⬇

Consider Quantitative Basis
(including uncertainties)

⬇

Identify Important Parameters, Models, and
Assumptions

⬇

Consider Evidence/Confirmation

Examples

- ◆ Illustrative of Concept
  - engineered system
  - natural system

- ◆ Examples are not regulatory
  requirements nor do they imply
  regulatory acceptance

Spent Nuclear Fuel Dissolution
- Identify Risk Significance -

◆ Risk Insights baseline indicates that spent nuclear fuel dissolution is risk significant:

"The dissolution of the waste form in an aqueous environment is important for all radionuclides. Uncertainty in the dissolution is large such that the time required to release radionuclides from the spent fuel matrix can vary from hundreds of years to hundreds of thousands of years."

08/13/2003

7

Spent Nuclear Fuel Dissolution
- Consider Quantitative Basis -

◆ Existing information has been used to develop models in the TPA code

◆ Four different models in TPA for dissolution of spent nuclear fuel
  - carbonate solutions (model 1)
  - presence of Si and Ca ions (basecase)
  - natural analog
  - secondary mineral formation (Schoepite)

08/13/2003

8

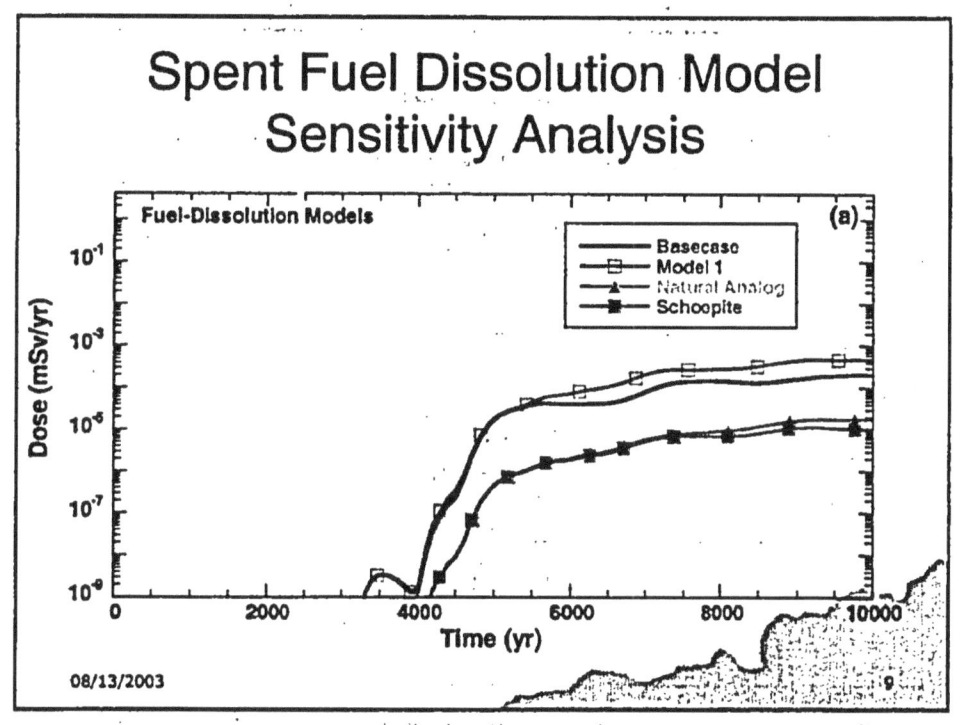

Spent Fuel Dissolution Model Sensitivity Analysis

08/13/2003

9

Spent Nuclear Fuel Dissolution
- Potential Importance -

◆ Limitations of the models were considered in developing risk insights baseline

◆ Parameter uncertainty
  - dissolution rate

◆ Model uncertainty
  - water chemistry
  - secondary mineral formation

08/13/2003

10

Spent Nuclear Fuel Dissolution
- Consider Evidence/Confirmation -

Dissolution Rate (mg/m2-day)	Sample	Solution (pH)	Test Method	Reference
0.2 - 1.0 ~ 1/140 for partially clad fuel	Spent fuel	J-13 (8.4)	Immersion	Wilson, 1990
3×10^{-2} − 3.0	UO$_2$	NaHCO$_3$ + CaCl$_2$ +Silicic Acid (8.4)	Flow Through	Gray and Wilson, 1995
(0.8 - 2.5) x 10^{-2}	UO$_2$	Silicate Solution (Near Neutral)	Flow Through	Tait, 1997
0.07 36 (initial, will decrease)	Spent fuel	Allard Synthetic Groundwater (8.1) (2.0)	Immersion	Forsyth, 1997
2.7	Spent fuel	J-13 (8.4, down to 3.2)	Drip	ANL, Finch et al., 1999
10 ~1/30 at pH 8 compared to pH 3	UO$_2$	HCO$_3$ (3) Reducing	Flow Through	Bruno et al., 1991

08/13/2003

11

Retardation in Alluvium
- Identify Risk Significance -

◆ Risk insights baseline indicates that retardation in the alluvium is risk significant:

"Retardation in the alluvium has the potential to delay the movement of most radionuclides for very long time periods (e.g., thousands to tens of thousands of years and longer) for nuclides that tend to sorb onto porous materials (e.g., Np-237, Am-241, Pu-240)."

08/13/2003

12

Retardation in Alluvium
- Consider Quantitative Basis -

- ◆ Existing information has been used to develop retardation factors for the TPA code
- ◆ Information for specific radionuclides
  - crushed tuff analog
  - literature values
- ◆ Support for conceptual model
  - linear isotherm
  - fast and reversible sorption reaction

08/13/2003

13

Retardation in Alluvium
Sensitivity Analysis
[years for initial release into Sat zone to exit Sat zone]

Nuclide	Alluv(1km) Rf (low)	Alluv(1km) Rf (high)	Alluv (5km) Rf (low)	Alluv(5km) Rf (high)
Tc 99	350	350	550	550
I 129	350	350	550	550
Np 237	950	76,000	1,050	>100K
Am 241	>100K	>100K	>100K	>100K
Pu 240	54,000	>100K	>100K	>100K

08/13/2003

14

Retardation in Alluvium
- Potential Importance -

- ◆ Extent of uncertainty
  - zero Kd (e.g., I and Tc)
  - range of Kd unimportant (e.g., Am)
  - range of Kd significant (e.g., Np)
- ◆ Sorption reaction is fast and reversible
- ◆ Changes in the bulk chemistry along the transport path

08/13/2003   15

Retardation in Alluvium
- Consider Evidence/Confirmation -

- ◆ Mineralogy of alluvium

- ◆ Water chemistry in alluvium (e.g., pH, ionic strength)

- ◆ Sorption Coefficient for Np
  - site-specific batch sorption tests
  - dynamic tests (flow-through column tests)

08/13/2003   16

Summary

- Risk insights identify areas of consideration for performance confirmation
- Uncertainties in parameters and models help determine extent of performance confirmation
- "Evidence" based approach
- NRC staff recognizes that DOE may make modeling selections (abstractions) that limit the significance of particular models and parameters

08/13/2003

17

Status

- Risk insights report to be completed in the October time-frame
    - based on risk baseline
    - provides quantitative basis
    - identifies further calculations

- Risk insights report will be updated as appropriate

08/13/2003

18

Left blank intentionally.

I.6 NRC's Acceptance Criteria in the Yucca Mountain Review Plan (YMRP) for Review of Performance Confirmation
Performance Confirmation Program - Section 2.4 of the Yucca Mountain Review Plan
Jeffrey Pohle, U.S. Nuclear Regulatory Commission

Jeffrey Pohle gave a talk about the portion of NRC's Yucca Mountain Review Plan that relates to performance confirmation. He noted that the review plan has four primary review areas. The first area consists of the general requirements, including the objectives to acquire data to indicate whether subsurface conditions are within assumed limits and whether barriers are functioning as anticipated. This first area also includes schedules, provision of baseline information, and the monitoring and analysis of possible deviations from baseline.

The second area of review deals with the confirmation of geotechnical and design parameters, such as the insitu monitoring of the thermomechanical response of the underground facility until permanent closure. The third review area addresses design testing, including, for example, tests of borehole or shaft seals and drip shields and the evaluation of thermal interactions of engineered barriers with the natural environment. The fourth review area concerns the monitoring and testing of waste packages.

Mr. Pohle noted that to achieve an adequate review of performance confirmation, NRC reviewers will need to be familiar with:

- barriers important to waste isolation and any unresolved NRC concerns
- DOE's description of the capability of each barrier to isolate waste
- DOE's information on uncertainties related to parameters, processes, models, etc., for each barrier
- DOE risk evaluations and NRC's risk insights baseline
- CNWRA support to enhance independent review capability.

Mr. Pohle stated that NRC needs an educated staff that is knowledgeable about DOE's description of what the barriers are, what the capabilities for the barriers are, the outstanding concerns in these areas, information about uncertainties, the evidence related to these parameters, and information from NRC-generated risk evaluations. Support from NRC's technical assistance contractor, the Center for Nuclear Waste Regulatory Analyses (CNWRA), will be needed to help enhance NRC's capability to independently review performance confirmation. CNWRA is currently doing work in the area of instrumentation, looking ahead at the types of testing activities DOE may propose to do and the instrumentation required. CNWRA is also looking at longer-term tasks on software requirements for future changes in computer codes, particularly thermohydrologic codes.

John Kessler, Electric Power Research Institute (EPRI), observed that there seems to be a disconnect between what NRC is emphasizing in performance confirmation and "almost everything else." He heard from NRC speakers an emphasis on every barrier, regardless of its individual contribution to overall performance. If DOE calls something a barrier, it appears NRC is going to ask them to defend it equally, whether it is the waste package or whether it is the saturated zone. Dr. Kessler noted that DOE considers the saturated zone to be relatively unimportant, but NRC considers it to be important. It appears that the two organizations are taking fundamentally different approaches, and this relates not only to performance confirmation but to the whole license application.

Mr. McCartin responded that NRC is looking at the potential to contribute to overall risk. For example, neptunium tends to be the largest dose contributor. In the natural system, the alluvium has the potential to significantly retard the most important radionuclide for overall risk. And that's why, with regard to neptunium, the saturated zone, and specifically alluvium, is important.

PERFORMANCE CONFIRMATION
PROGRAM
Section 2.4 of Yucca Mountain
Review Plan

144th Meeting of
Advisory Committee on Nuclear Waste
July 29-31, 2003

Jeffrey Pohle 301-415-6703 jap2@nrc.gov
Division of Waste Management
U.S. Nuclear Regulatory Commission

Discussion Topics

➢ **Overview of Section 2.4 of Yucca Mountain Review Plan in terms of the four primary areas of review**

➢ **NRC reviewer's information needs**

YMRP Section 2.4 Overview

Areas of Review

➢ General requirements including:
  ➢ Objectives to acquire data by identified tests to indicate whether subsurface conditions are within limits assumed in licensing review and whether natural and engineered barriers are functioning as anticipated
  ➢ Overall schedule
  ➢ Implementation with regards to adverse effects of program, provision of baseline information, and monitoring and analyzing changes from baseline

YMRP Section 2.4 Overview

Areas of Review (continued)

➢ Confirmation of geotechnical and design parameters including:
  ➢ Measuring, testing, and mapping during construction and operation to confirm geotechnical and design parameters related to natural barriers
  ➢ Monitoring, in situ, the thermomechanical response of the underground facility until permanent closure
  ➢ Surveillance program to evaluate subsurface conditions against design assumptions including provisions for comparing observations with design bases and assumptions, determining need for changes to design or construction methods, and reporting comparative differences, their significance to health and safety, and recommended changes, to the Commission

YMRP Section 2.4 Overview

Areas of Review (continued)

➢ Design testing Including:
   - ➢ Testing of engineered systems and components, other than waste packages, used in the design (for example, borehole or shaft seals, drip shields)
   - ➢ Program to evaluate thermal Interaction effects of waste packages, backfill, drip shields, rock, and unsaturated zone and saturated zone water
   - ➢ Plan to test, before permanent placement begins, effectiveness of backfill placement and compaction procedures against design requirements (if backfill is used)
   - ➢ Plan for tests to evaluate effectiveness of borehole, shaft, and ramp seals before full-scale sealing begins

YMRP Section 2.4 Overview

Areas of Review (continued)

➢ Monitoring and testing waste packages Including:
   - ➢ Plan for monitoring the condition of waste packages at the geologic repository operations area, including an evaluation of the representativeness of those waste packages chosen for monitoring and representativeness of the waste package environment of waste packages chosen for monitoring
   - ➢ Plan for laboratory experiments that focus on the Internal conditon of waste packages, including evaluation of degree environment within underground facility duplicated in laboratory
   - ➢ Duration of the waste package monitoring and testing program

Performance Confirmation Plan Review

To achieve an adequate review context and focus, NRC reviewer's need to be familiar with:

➢ Barriers important to waste isolation identified by DOE (and any outstanding NRC concerns)

➢ DOE's description of the capability of each barrier to isolate waste (and any outstanding NRC concerns)

➢ DOE's information on uncertainties related to parameters, processes, models, etc. relevant to individual barrier's waste isolation capability

➢ Available DOE risk evaluations

➢ NRC's risk insights baseline

➢ CNWRA support to enhance independent review capability

128

5. STAKEHOLDER PRESENTATIONS

Left blank intentionally.

5.1 Nye County's Views on Performance Confirmation and Related Topics
Les Bradshaw, Nye County, Nevada

Les Bradshaw (Nye County, Nevada) gave a talk titled "Nye County's Views on Performance Confirmation and Related Topics." He noted that performance confirmation is a critical program element because it will show whether the repository will perform in a way that protects health and safety in Nye County. Mr. Bradshaw was concerned that no approved program appeared to be in place. He expressed concern that DOE suspended monitoring of several unsaturated zone boreholes in 2001. He felt that this monitoring should have been a part of a performance confirmation program.

Mr. Bradshaw considers that a comprehensive performance confirmation program should have been in place long ago and that Nye County, Nevada, and other stakeholders should have had a chance to review it. Confirmatory studies should include significant participation by qualified groups from outside of DOE. Performance confirmation will be more acceptable to the public if some of the work is done by qualified independent groups. The following tasks could be undertaken by independent entities: (1) technical review of plans, data, and analyses, (2) establishment of baseline data for water, air, rock and soil, and biota, (3) post-emplacement monitoring of the environment; and (4) storage and dissemination of performance confirmation data. Nye County is already participating in performance confirmation work that will be related to Nye County's Early Warning Drilling Program. DOE has approved funding for expansion of this work through 2007.

Left blank intentionally.

Nye County Department
of Natural Resources and Federal Facilities

Nye County's Views on Performance Confirmation and Related Topics

Presented by:
Les Bradshaw

ACNW Working Group Session on Performance Confirmation Plans
July 30, 2003

Introduction

- Nye County has always considered Performance Confirmation (PC) as a critical program element because it will demonstrate whether the repository will perform in a manner that protects the human health, safety, and the environment in Nye County.

2

Nye County Department of Natural Resources and Federal Facilities

Regulatory Requirements

- Under 10 CFR 63.131(b) the Performance Confirmation program must have been started during site characterization.
  - Has it?
  - No approved program is in place to our knowledge.

3

Nye County Department of Natural Resources and Federal Facilities

Regulatory Requirements (Continued)

- 63.131(d) (2) provides that PC program must be implemented so that it provides baseline information and analysis of that information on those parameters and natural processes pertaining to the geologic setting "that may be changed by site characterization, construction and operational activities".
  - Has this requirement been met?
  - What about the DOE decision to suspend monitoring of UZ boreholes in 2001? This should be a component of any PC program.
  - Nye offered to pick up that effort, but DOE turned down.

4

Nye County Department of Natural Resources and Federal Facilities

Regulatory Requirements (Continued)

- Under 63.131-134 all facets of the repository must be subject to a PC program.

- In summary, a comprehensive PC program should have been in place, or at least designed and subjected to independent stakeholder review and input, long ago.

5

Participation in PC by Independents

- Nye and many others believe that PC should include significant participation by qualified organizations that are independent of DOE.
- PC performed by DOE will not be as acceptable to the public as PC performed by qualified independent entities.
- PC tasks that could be conducted by these independent organizations include, but are not limited to:
  - Technical review of plans, data, analyses, and interpretations beyond the NRC licensing process.
  - Establishment of baseline data for environmental media including surface and subsurface water, air, rock/soil, and biota.
  - Long-term monitoring of environmental media beginning when waste is first received (i.e. post-emplacement monitoring).
  - Storage and dissemination of PC data.

6

Nye's Present PC Capabilities

- Nye, via its successful Independent Scientific Investigations Program, is presently participating in, or positioned to participate in, a number of PC tasks. For example:
- Nye's Early Warning Drilling Program has and continues to demonstrate technical expertise in establishing and operating a groundwater monitoring network downgradient from Yucca Mountain.
    - Nye currently collects and analyzes groundwater samples and water levels from this network for independent baseline monitoring and shares samples and data with DOE and NV.
    - This network, with Nye as the qualified operator, should serve as the basis for post-emplacement groundwater monitoring downgradient from Yucca Mountain.
    - DOE has in principle approved funding for the continued expansion of this downgradient network through 2007.

Nye County Department of Natural Resources and Federal Facilities

---

Nye's Present PC Capabilities (Continued)

- Nye is well qualified to extend and operate this downgradient groundwater network to adjacent regions on the Nevada Test Site that may be impacted from nuclear testing.

- Nye presently employs/contracts a group of technical experts in subsurface hydrogeologic characterization and monitoring who are well qualified to:
    - Independently review PC plans, data, and analyses.
    - Plan and conduct vadose zone PC monitoring of air, water, and rock in the repository and in surrounding boreholes.

Nye County Department of Natural Resources and Federal Facilities

Nye's Plans for Developing Additional PC Capabilities

- Nye is working towards developing in the near future the expertise, organization, and facilities to participate in other PC tasks suitable for independents including:
  - PC monitoring of surficial environmental media (soil, air, and biota) downgradient from Yucca Mountain as well as within adjacent regions of the Nevada Test Site that may be impacted from nuclear testing.
  - Storage and dissemination of data.

9

Nye's Plans for Developing R&D and Operational Related Capabilities

- Nye is also working towards developing the capability of managing and hosting other Yucca Mountain related development, manufacturing, and construction activities including:
  - Development of instrument systems for remote monitoring of subsurface conditions in the repository and in monitor wells or boreholes.
  - Manufacturing waste cask prototypes and production units.
  - Construction of facilities necessary to support training, monitoring, sample archival, and data storage and dissemination.

10

PC vs. Research and Development

- There seems to be some confusion today about the difference between long-term R&D and PC

- PC should be considered to be those scientific activities, including long-term monitoring, that assures the repository is, or likely will, operate as expected, and that thus assures license compliance.

- R&D should be other scientific investigations designed to enhance understanding of the system, both natural and engineered, and that might be used to improve repository performance in the future.

- PC is linked to, but separate from R&D
  - e.g. As proven cost-effective R&D advances in monitoring become available they should be incorporated into PC.

11

Nye County Department of Natural Resources and Federal Facilities

Planning and Budgeting for PC and R&D

- PC and R&D are both complex and long-term undertakings. Neither in the past have been included in the routine budgetary process, or in the Total System Life Cycle Cost (TSLCC) report, except in very general terms.

- How one prioritizes and funds PC and long-term R&D is not yet clear. It is essential that we start focusing on this now, rather than cob together some program at the last minute for the LA that no one else has had a chance to provide input.

- When you look at some examples of the activities involved in PC that will probably be most difficult (remote monitoring of waste packages, e.g.) it becomes clear that budgetary considerations and decisions other than purely scientific ones will be important, if not the drivers.

12

Nye County Department of Natural Resources and Federal Facilities

Planning and Budgeting for PC and R&D (Continued)

- The institutional arrangements for conducting PC and R&D over the very long-term have not been examined.
    - i.e. Nye would like to be involved in the development of instrument systems for remote monitoring.
- Confidence in the long-term stability of the independent organizations involved, not just DOE, will also be critical.
- Nye studies indicate that current fee structure may be inadequate to fund the DOE TSLCC, even without adding in long-term R&D and PC costs.
- No consideration has yet been given to the resolve of Congress to continue appropriating large sums of money once spent fuel and high-level waste is "out-of-sight and out-of-mind".

13

Nye County Department of Natural Resources and Federal Facilities

---

Summary

- DOE has much PC ground to make up.
- PC and R&D are different and detailed planning and budgeting should be completed ASAP.
- Qualified independent entities should be involved in PC.
    - e.g. Similar to the role the Electric Power Research Institute (EPRI) has undertaken in addressing PC strategies and developing a prototype PC plan.
- Nye has unique qualifications and should play an active role in PC and R&D.

14

Nye County Department of Natural Resources and Federal Facilities

Left blank intentionally.

5.2 Some Observations on Performance Confirmation and Performance Assessment
John Walton, University of Texas at El Paso

John Walton (University of Texas at El Paso, consultant to Nye County) gave a talk titled "Some observations on performance confirmation and performance assessment." Nye County has several areas of concern, including the anticipated impacts of a repository on Nye County resources and potential unresolved performance assessment issues. Dr. Walton suggested that the future heating up of the mountain will cause the top of the mountain to become warmer and wetter, resulting in possible changes in flora and fauna. These changes could take place in tens to hundreds of years. Therefore, soil conditions and vegetation changes should be monitored over time. A baseline of vegetation communities should be obtained before a repository is built.

Dr. Walton observed that tunnel roof collapse remains an unresolved question, because rubble would act as insulation and change conditions assumed in coupled thermo-hydrologic modeling. Backfill may be needed to provide a predictable environment. Dr. Walton was concerned that the natural ventilation of the mountain may not be fully accounted for in DOE models. This is important for heat, moisture, and chemistry modeling. He also stated that DOE's models mix spatial and temporal variability with uncertainty, which can unrealistically spread projected risk and reduce peaks in mean projected doses. Dr. Walton wonders if this mixing of variability and uncertainty is conservative or nonconservative in the context of Yucca Mountain.

Dr. Walton gave an example of a simplified "pseudo" performance assessment that included four processes: corrosion, release rate, transport lag time, and an unspecified event that fails the remaining waste containers when it occurs. He compared two simulations. In one simulation he took the mean dose representing 1,000 Monte Carlo realizations. The results are compared to a second simulation, where the standard deviation is increased for one parameter, which increases the uncertainty range. Contrary to expectation, in this latter case the risk is actually reduced because it is measured as the peak of the mean of the realizations. What happens is, sometimes when you modify a parameter, each of the 1,000 realizations will have its peak occur at different points in time. That is, the peaks of the individual realizations will be spread in time. So when the mean is calculated, it broadens and flattens relative to the curve with less variance. The projected risk is lower, and performance has actually been improved by ignorance. This is not a general conclusion, because if different parameters are changed, sometimes the risk increases. The results depend on which parameter is broadened. It's complicated and not obvious what the result will be. Therefore, in performance assessments for Yucca Mountain, when are so-called "one-off" and "one-on" analyses conservative or nonconservative? Dr. Walton described a scenario in which a DOE manager is asked to fund a study on the sorption coefficient (K_d) of neptunium. Will the manager really want to fund it if credit is being taken for the fact that the K_d isn't well known? In conclusion, Dr. Walton noted that local involvement is crucial to performance confirmation because otherwise the work is the product of an internal "group think" and doesn't produce as many ideas. Dr. Walton stated that Nye County should be involved.

Left blank intentionally.

Nye County Department
of Natural Resources and Federal Facilities

Some Observations on Performance Confirmation and Performance Assessment

John Walton
University of Texas at El Paso

ACNW
July 2003

Areas of Concern

- Monitoring of anticipated impacts on Nye County resources
- Unresolved performance assessment issues

Anticipated Impacts

- Heating of mountain and induced airflow at YM
  - dryer and cooler below
  - warmer and wetter above

  condensation

  evaporation

  - will induce flora and fauna changes

Sequence of Events

- Mountain heats up

- Increased natural breathing of mountain

- Changes to flora and fauna on scale of 10's to 100's of years

- Monitor soil conditions and vegetation changes

- Adequate pre construction vegetation analysis necessary for baseline

Unresolved Questions

- Roof collapse
  - many analysts anticipate roof collapse in 10s to 100s of years
  - DOE modelers assume drifts are eternally open
  - rubble makes good insulation
  - THC modeling is of limited utility if we don't know the "R-value" in the attic
  - if the situation is uncertain, backfill may be required to provide a predictable environment

Nye County Department of Natural Resources and Federal Facilities

Unresolved Questions

- Extent of natural ventilation
  - Repository will increase natural breathing of mountain
  - Not fully in DOE models
  - Important for heat, moisture, chemistry modeling

Nye County Department of Natural Resources and Federal Facilities

Unresolved Questions

- Uncertainty vs. variability
  - By necessity performance assessment models mix spatial and temporal variability with uncertainty
  - This can lead to unrealistic spreading (dilution) of projected risk, thereby reducing peaks in the mean projected dose curve
  - Mixing of variability and uncertainty is not realistic, but
    . . .
  - In YM context is it conservative or non conservative?

Example Calculations – a simplified PA

- Processes:
  - Corrosion
  - Release
  - Transport
  - Event
- Arbitrary units, 1000 realizations
- Normally distributed parameters
- If we assume we are God for a moment, we can run the calculation both ways

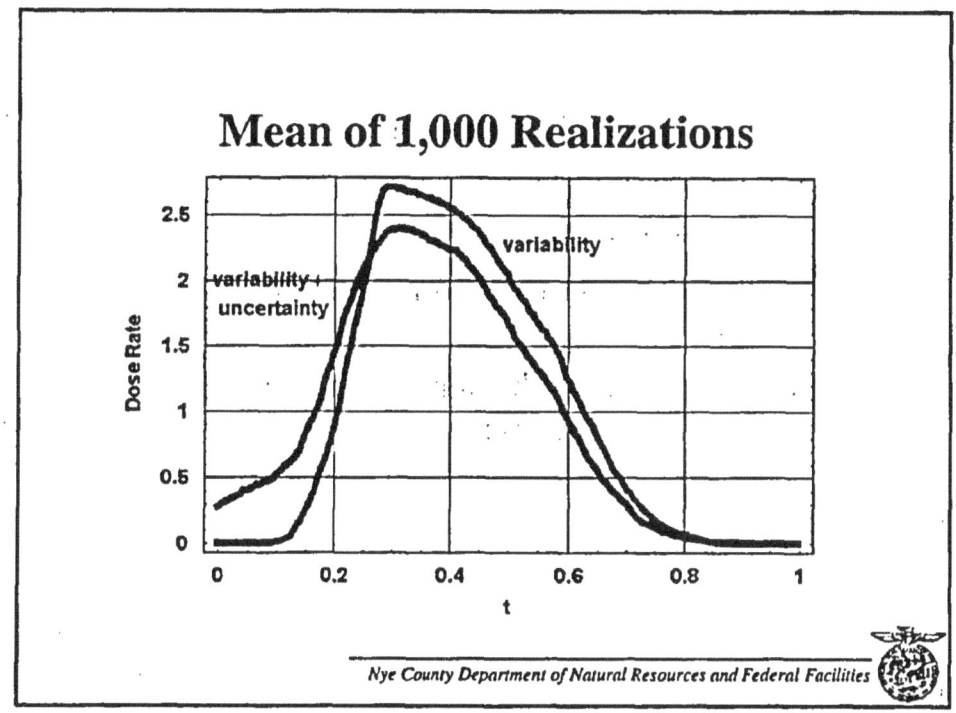

Mean of 1,000 Realizations

Result

- Inclusion of uncertainty reduced projected risk as measured by the "peak of the mean of the realizations"
- Sometimes inclusion of uncertainty increases projected risk
- The difficulty is caused by the metric, "peak of the mean of the realizations"
- With this metric, inclusion of uncertainty may either increase or decrease projected risk in a difficult to discern pattern
- What does it do in TSPA?
- What incentive does DOE have to reduce uncertainty when it can increase projected performance?

Conclusion

- Many important issues remain
- Local involvement in performance confirmation is essential
- Nye County can work cooperatively to help resolve some of the issues

Nye County Department of Natural Resources and Federal Facilities

148

Commentary by Representative from the State of Nevada
   Steve Frishman, State of Nevada

/IR. Steve Frishman: As you notice, I did what I have often done with ACNW working groups)efore, and that is I don't commit anything to paper because I think the purpose of the working jroup is to try to work through issues and topics and not just have paper to walk away with and ;ay, "Okay. We have our stack of paper for today."

n the last day and a half, we've tripped over I think most of the obvious questions that are out here about performance confirmation that we have all, in one way or another, talked about over ι number of years. One point to remember is that this is nothing new to 10 CFR Part 63. The)erformance confirmation requirement is essentially identical to the requirement that was in ι0 CFR Part 60. Its meaning hasn't changed either from what I can tell. Also, it appears to me hat performance confirmation has been analyzed out of the regulation by the review plan.

;o I am not sure there is a lot to do about a further understanding of performance confirmation n the sense of looking to the Commission to maybe reinterpret or further interpret. It's sort of here, but we still have this big question, what is performance confirmation in terms of the /arious interests from both the applicant side and from the regulatory side and, of course, from he review side ultimately? We have to remember, first of all, what performance confirmation is ;aid to be in the rule. I noticed that nobody in the last day and a half has actually gone back to he definition of performance confirmation.

t's probably instructive to remember that it says—this is not verbatim, but this has stuck in my nind for a long time—performance confirmation is a program to confirm the validity of the nformation used to demonstrate and support the reasonable expectation determination. As vas mentioned yesterday and again today, performance confirmation is to begin during site ;haracterization and continue through repository closure.

;o let's think about what the real purpose of performance confirmation must be. f you put it in the context of the regulatory process, it seems like its purpose is relatively ;imple—to provide additional confidence in the technical basis for a decision to amend the icense for repository closure.

t is probably important to keep it in that context. The reason for that is a discussion that \CNW and others with the Commission and other places have heard from me before—under he regulation, the disposal decision is made with the construction authorization decision. And ιll changes after that are amendments in one way or another, but they need support that)riginal disposal decision.

/Vhat I see performance confirmation inching towards, even though there are statements to the ;ontrary, is as Chris Whipple put it yesterday, a kind of "bucket." I see a danger of unfinished)usiness in site characterization being casually flipped into this "bucket" of performance ;onfirmation. And, in fact, I had a thought. When Tim McCartin was doing his presentation oday, where if you look at his presentation and just do a few minor word changes here and here, the title really should be "Risk-Informing Performance Assessment." McCartin picked a ;ouple of narrow examples of how to do that. So we are in a situation where it is pretty clear hat there are a number of areas where site characterization is not complete. But, at the same time, there is the recognition that the license application has to be one that is adequate for a decision regarding reasonable expectation that the performance requirements will be met.

Because of the circumstances of this program, we are in this sort of push/pull. I would be greatly concerned if there was any approach literally on the part of anyone to try to use performance confirmation to overcome this incomplete site characterization and actually get to a point where it gains significance in licensing. Now, probably the key message here is that the license application review and the hearing should proceed to a reasonable expectation decision without any deference whatsoever to the substantive content of the performance confirmation program. Performance confirmation is essentially an add-on. And it should have literally no basis in the disposal decision that comes at the time of a decision on construction authorization. Yes, it's a good thing to do. And it is a good thing to do for a couple of reasons that I want to get into. But it should be, as I said, given no deference, meaning that yesterday's comment from Jim Blink towards the end was certainly a friendly offer from the standpoint of making things operationally simpler, but it also was sort of a violation of this because what he invited in one of the tough spots was, "Make it a license condition." What I see coming is making a lot of things into license conditions and license conditions hooked into this vehicle or bucket of performance confirmation so that we get in that situation where site characterization is never ending.

We know that performance assessment will go on forever, as it probably should. But that first performance assessment had better be demonstrably good enough in every possible way. So the performance confirmation program itself may be looked at in a light a bit different from the direction that I think the staff is going with its risk-informing, a little bit different perhaps from the way Chris Whipple was describing in terms of "pick out what is most important and go after that." I think there are two things going on. One of them is yes, it is very important to look at the things that are most important, but it's also very important to have a place for the necessary ongoing baseline data collection. Baseline is important because if the repository goes forward at all, you are going to have people doing construction and disturbing things for many, many years.

And the rainfall discussion yesterday was a good one. What do you do if the rain falls out of compliance? It should not be a difficult question because there shouldn't be a question of whether the rainfall is in compliance. But what it does is it drops things into sort of two boxes. One box consists of the things that are most important, and how do we get at them, remembering all the time that further major discoveries are most likely to be adverse, rather than in your favor. Things just seem to happen this way. So we can't get in a situation where you can say that we're looking for good things in the future to make up for what we don't know now. You can't do that. And I have told the NAS committee on staging the same thing.

You can't set up a situation where you expect good things to help you out of what may be just marginal right now. The future isn't going to bring you that unless you are really lucky. More likely it will bring you things you don't want to know, rather than things you do want to know. So looking at the things most important to risk, yes, that is necessary to do because you are in a situation where information will be made available throughout this long period of time and information that, of course, is important to what you think now about performance. There is also a lot of other information that I think the performance confirmation requirement gave an incentive to collecting. And that's just the ongoing information that is available, such as weather, such as you've only got 5 miles of tunnel right now or 6 miles, where only a small portion of it is in what the current design shows will be the vast majority of the emplacement

150

ock. If this all goes forward, it's going to be another up to about 100 miles of tunnel in that rock over a horizontal space that is known to vary from north to south anyway. And there is data that needs to be collected that we could call confirmatory, I think, if that is a regulatory word we are going to use. But what it is intended to tell you is if you collect it properly, that rock has properties that either are or are not within the ranges that were anticipated in the models. This is as a matter of course the type of thing that should be done.

There was a question earlier today about "as anticipated." Well, what is anticipated right now for the underground comes from data that has been collected in a pretty small place compared to the larger area that could be excavated. "As anticipated" in this case means you look at all of it to make sure its hydrologic properties are within the range that your models were based on. Chances are you will find things that are not within that range. And then what do you do about it? That needs to be, as someone said yesterday, in the pre-thinking "What do you do about it?" as opposed to the post-thinking "What do you do about it?" because we have a myriad of examples in this program where the answer to "What do you do about it?" is go out to prove that it doesn't matter. And if you think about it ahead of time, that is not your first natural reaction over what you would do about new information.

What I am urging is that performance confirmation be seen as an organized way to collect new data about underground conditions during the construction of new drifts and tunnels. Also, performance confirmation should take a very hard look at the performance approach that has been taken and should not think so much in terms of what is most important, not do endless reiterations and rethinking about the components of the waste package model. The most important thing is to go back, look at, and challenge the conceptual models on which the performance assessment was built.

If you will remember, it is less than 10 years ago that a monstrous change in the conceptual model of a Yucca Mountain repository had to be made. And it was not expected 12 years ago, but starting about 10 years ago, it was essentially mandatory that it be made. It's not unlikely that additional data are going to lead to the necessity to make other analyses of whether the conceptual models behind performance assessment are sufficiently representative to be carried forward. So what I am saying is that performance confirmation allows a framework to do something that I think would be totally inappropriate, which is be a bucket for everything that is undone. It also invites something much more rational, which is a way of dealing in an organized way with a common sense data flow that comes from the ongoing activity. Performance confirmation can provide information to challenge the real basis of safety, which is a short string of conceptual models that have led to a decision that would allow you to dig these extra tunnels in the first place, if there is even enough information for that.

So my caution is that you don't use this workshop and all the presentations that have been made to try to revisit what performance confirmation could be if it were to be most friendly to a license application, most friendly to the applicant, or maybe even most utilitarian to the regulator. Performance confirmation is a pretty simple thing to be used in a common sense way, not in a way that results in an uncertain job only becoming more uncertain because someone found it to be a convenient way because it is the only bucket left out there to throw stuff into.

Left blank intentionally.

5.4 Commentary by Representative from the Las Vegas Paiute Tribe
Atef Elzeftawy

Atef Elzeftawy (consultant) presented remarks to the Committee on behalf of the Las Vegas Paiute Tribe. He discussed a number of topics, including the need for regulators to be tenacious in finding out how DOE plans to do the work. He had discovered years earlier that DOE was planning to drill unsaturated zone holes using drilling mud. That wouldn't have worked because the mud would have infiltrated the pores and fractures of the rock along the boreholes, interfering with the ability to measure hydrologic parameters for the unsaturated zone. DOE changed its methods. He emphasized the importance of unsaturated zone hydraulic parameters. Dr. Elzeftawy also suggested that the tribe had a concern that funding for things like performance confirmation might initially grow, but later dwindle to almost nothing. He submitted to the Committee an article about the golf resort owned by the tribe. This article has been placed in the formal record of the meeting.

Left blank intentionally.

5.5 Performance Confirmation - What Does it Really Mean?
Engelbrecht von Tiesenhausen, Clark County, Nevada

Mr. Engelbrecht von Tiesenhausen (consultant to Clark County, Nevada) spoke on the topic "Performance Confirmation, what does it really mean?" He discussed the general requirements and definitions of performance confirmation that are described in NRC's Part 63 regulation. Mr. von Tiesenhausen noted that there are several challenges, such as estimating temperature effects in a repository and the idea that even in a dedicated drift for performance confirmation, conditions are unlikely to reproduce those found in a repository. Clark County considers waste package performance to be the most critical performance issue. Long-term corrosion data in a representative environment is "most likely impossible" to collect before a repository would be closed. Mr. von Tiesenhausen stated that performance confirmation should not be used to put off the resolution of issues that are part of a license application. The program should confirm results but not be a primary source of data. Any license application that relies on performance confirmation and formal "requests for additional information" "should be looked at very critically." Mr. von Tiesenhausen suggested that confirmatory studies can help us better understand the natural system in several ways. For example, such work can improve the understanding of the role of the Calico Hills geologic formation on waste isolation. It can help to better interpret where and how fast water travels in the natural system. And finally, confirmatory studies can improve the understanding of current and future geochemical processes.

Left blank intentionally.

PC, what does it really mean?

Comments by
Clark County to the ACNW

1

Department of Energy Statement

- The Strategy of the Performance Confirmation program is to utilize multiple data acquisition methods to produce an overall data set which is adequate to **confirm (or revise)** licensing assumptions about repository performance.

2

Nuclear Regulatory Commission 63.131

- (a) The PC program must provide data that indicate, where practible, that (not direct quotes)

  (1) Actual subsurface conditions are within the limits assumed

  (2) Natural and engineered systems are functioning **as intended**

  (b) The program must have been started during site characterization and it will continue until permanent closure.

3

Definitions Cont.

- 63.103(M) Performance Confirmation

  A performance confirmation program will be conducted to **evaluate the adequacy** of assumptions, data, and analyses that led to the findings that permitted construction of the repository and subsequent emplacement of the wastes.

4

EPRI Report on Performance Confirmation

- any decision by the NRC to license each stage of repository development, if a license application should be tendered, would be made on the basis of the information that **exists at the time** that the NRC considers such an application.

5

Challenges

- Temperature effects are difficult, if not impossible to scale.
- In processes that are well understood the effects of long time periods can be compensated for by changing other independent variables
- Even in a dedicated drift for PC, conditions are unlikely to duplicate those in the repository.
- Some of this data will still be useful.

6

Concerns

- Waste package performance is still the most critical issue from a performance standpoint

- Data on long term corrosion in a representative repository environment is most likely impossible to collect prior to closure

- Data collected during the PC period should not be used to close agreements, or to be the primary data for TSPA for LA.

7

Current Schedule

	CLST	ENFE	IA	PRE	RDTME	RT	SDS	TEF	TSPAI	USFIC	Total
7-12/03	9	20	9	3	7	14	0	2	15	11	90
1-6/04	14	4	1	0	2	9	6	5	23	3	67
7-12/04	12	4	0	2	14	2	0	1	13	1	49
1-6/05	2	0	0	0	0	0	0	0	0	2	4

8

PC

- Should not be used as to means to defer the resolution of issues that are part of LA
- Should **confirm**, but not be the primary source of data
- An LA that relies on PC and RAI's should be looked at **very critically**.

9

PC or ST
Understanding the Natural System

- Improve the understanding of the role of the Calico Hills on waste isolation
- Water, where does it go and how fast does it get there?
- Geochemistry, current and future?

10

Left blank intentionally.

5.6 The Role of Performance Confirmation in Yucca Mountain Development
John Kessler, Electric Power Research Institute

Dr. John Kessler gave a talk titled "The role of performance confirmation in Yucca Mountain development." He described differences between performance confirmation and long-term research and development. PC is specifically designed to evaluate the technical bases for the licensing decision. EPRI has performed several activities related to PC, including evaluation of DOE's draft 2000 report, convening of a PC panel to make recommendations, hosting of a performance confirmation workshop in 2001, and documentation of the above in a December 2001 EPRI report. Dr. Kessler recommended that NRC and DOE start now to develop a shared understanding of how both performance confirmation and long-term monitoring will be carried out. A flexible plan is needed, with work activities to be prioritized using risk-informed judgment. He noted that NRC and DOE have made a commendable start, NRC with its risk-informed regulation and DOE with an initial PC Plan.

Dr. Kessler described possible criteria for prioritizing activities, such as (1) risk-informed, (2) timing of the need for data, (3) cost of an activity, (4) interference with other activities, (5) stakeholder agreements, and stakeholder concerns, (6) health effects to workers and the public, and (7) ability to define the activity in such a way that "confidence" would be enhanced. Traps to be avoided include agreeing to measure things that do not affect performance, agreeing to do things that can't be done, requiring unnecessary accuracy or precision, monitoring for too short a time, assigning excessive levels of conservatism, and neglecting the need to maintain technical capabilities.

Dr. Kessler suggested a number of options to address important FEPs (features, events, and processes) that are not amenable to performance confirmation testing. These options are to (1) use reasonably bounding values based on expert elicitations, (2) leave a reasonable margin, (3) use natural analogues, and (4) add or modify engineered features to reduce the importance of the FEPs. These types of FEPs should be identified early.

Dr. Kessler advised that meaningful tolerance bands need to be established now, that a clear beginning and end must be defined for performance confirmation activities, that appropriate "baseline" information must be collected at the right times, and finally, that activities should be prioritized in case of limited funding or time.

Left blank intentionally.

The Role of Performance Confirmation In Yucca Mountain Development

John Kessler

Manager, HLW and Spent Fuel Management Program

Electric Power Research Institute

1-650-855-2069; Jkessler@epri.com

Presented to the NRC Advisory Committee on Nuclear Waste, 30 July 2003

EPRI

Background: Uncertainty is Unavoidable. How can it be "Managed"?

- Regulatory approaches:
    - Dose to a "reasonably maximally exposed individual"
    - RMEI dose limit a fraction of natural background
    - Multiple barriers
    - Waste must be retrievable
    - Long(er)-term R&D:
        - "Safety Questions" provision in NRC review plan
        - Performance Confirmation program
- Additional DOE approaches:
    - Reduce uncertainties with design modifications
    - Analyses conservative (on the whole)
    - "Margin": below, not at the limit
    - Long-term R&D / Performance Confirmation program

 2 EPRI

Distinction Between Long-Term R&D and 'Performance Confirmation'

- **Performance confirmation:**
  Activities that are specifically designed to evaluate the technical bases for the licensing decision

- **Long-term R&D:**
  Any other activity not specifically directed toward evaluating licensing bases

EPRI Work on Performance Confirmation (PC)

Work done in 2000-2001

- Evaluation of early (2000) DOE draft PC report
- Convene PC panel and make recommendations and observations
- PC workshop (DOE, NRC, NWTRB, Nevada counties, PC panelists, others), November 2001
- Provide examples of some appropriate PC activities using DOE "8-step" methodology

All of the above summarized in a December 20001 EPRI report (EPRI report number 1003032)

EPRI Performance Confirmation Panel Members

- Chris Whipple, Environ, Inc. (chair)
- Robert Budnitz, Future Resources Associates, Inc.
- Matthew Eyre, Exelon Corp.
- Barry Gordon, Structural Integrity Associates
- John Kessler, EPRI Inc.
- Rodney McCullum, Nuclear Energy Institute
- William Miller, QuantiSci Enviros, Ltd.
- Warner North, NorthWorks Inc.
- Alan Ross, Alan M. Ross and Associates
- Alice Shorett, Triangle Associates
- John Taylor, EPRI (retired)

EPRI

EPRI Performance Confirmation Panel December 2001 Comments

- PC (and other long-term R&D) is useful and appropriate
- There are many interested parties in PC
- NRC and DOE need to start now developing a shared understanding of how long-term R&D and PC will be carried out
  - Commitments will be identified in the license application and any near term amendments
- A flexible, adaptive plan is needed
  - Implications for using a rather rigid license amendment process?
- Prioritize now using risk-informed judgment and clear criteria for prioritization
- Avoid "traps"

EPRI

NRC and DOE Need Shared Understanding of PC/L-T R&D

- Commitments would likely be defined in the licensing process – even those not starting until much later
    - Concern is that DOE must get it right the first time, which is counter to a flexible, adaptive PC approach
- NRC and DOE have both made a commendable start
    - NRC: Final regulation
    - DOE: Draft PC and long-term R&D plans (Rev. 2 soon?)
- Differences between the two PC approaches need to be resolved
    - DOE: overall performance objectives are achieved
    - NRC: natural and engineered barriers are functioning as intended and anticipated

 7

  EPRI

Use Risk-Informed Judgment and Clear Criteria for Prioritization - Now

Potential criteria:

- The relative "value" of information (i.e., risk-informed)
- Timing of the need for specific information
- Cost of conducting a specific activity
- Interference with other activities
- Agreements with stakeholders
- Concerns of stakeholders
- Potential health effects to workers and the local population
- Ability to define sufficiently the activity such that "confidence" is truly enhanced in a reasonable amount of time

 8

EPRI

Traps to Avoid in Defining a Long-Term R&D Program

- Agreeing to measure parameters that do not affect performance
  - Satisfying parochial interests
- Agreeing to do things that can't be done
  - Requiring unnecessary accuracy or precision in measurements
  - Monitoring of too limited duration or extent
- Assigning excessive levels of conservatism on bounds because it is easy ("eats" margin)
- Neglecting institutional aspects (must maintain technical capabilities; periodic "report cards")

EPRI

DOE's Eight Steps* in Defining a Performance Confirmation Activity
(from DOE's 2000 draft PC report)

1. Identify which processes are to be measured, the 'key' performance confirmation factors [*DOE PC Rev. 2*]
2. Define data base and predict performance [*DOE PC Rev. 3*]
3. Establish tolerances or predicted limits or deviations from predicted values [*Rev. 3*]
4. Identify completion criteria and guidelines for corrective action [*Rev. 3?*]
5. Conduct detailed test planning [*Rev. 3*]
6. Monitor performance, perform tests, and collect data
7. Analyze and evaluate data
8. Recommend and implement appropriate actions if there are deviations [*discussed in Rev. 3?*]

   *The "steps" can be iterative

EPRI

Step 3. Establish Tolerances, Limits, Deviations from Predictions

- This is a key step in a successful performance confirmation activity

- Combine baseline data with predictions for performance confirmation period

- May become license conditions
  - i.e., "If...then" and "If not...then" specifications

11

EPRI

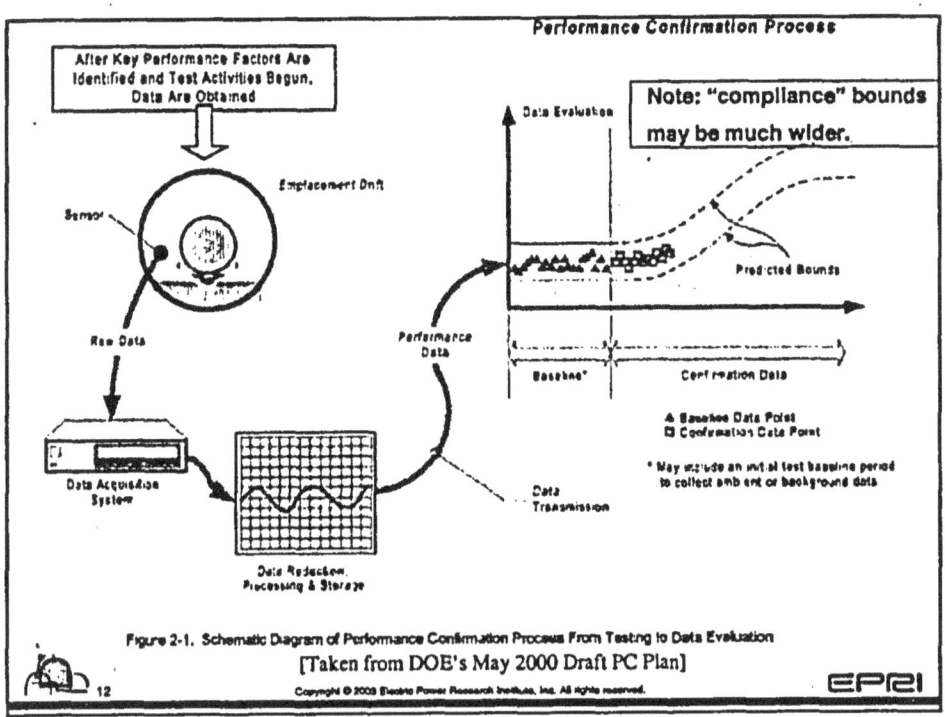

Figure 2-1. Schematic Diagram of Performance Confirmation Process From Testing to Data Evaluation
[Taken from DOE's May 2000 Draft PC Plan]

12

EPRI

Step 4. Identify Completion Criteria

- A clear end point must be identified
  - Tolerance bands at 50, 100, 200 years need to be developed
  - Test must be sensitive enough to detect the required tolerance
  - Test must be long enough
    - Need to know in advance adequate time is likely
    - Will be difficult to exactly define up-front how much time is required
- Sample size and frequency issues must be considered
  - E.g., must *every* container be examined?

 13 EPRI

Step 8. Recommend and Implement Appropriate Actions

Potential options:
- No action
- Limited, additional testing (if endpoint adequately defined)
- Modification of original licensing bases
- Engineered design modification(s)
- Temporary halt of emplacement
- Retrieval / abandonment of site

 14 EPRI

Suggested Options for Important FEPs not Amenable to PC Testing

- Use reasonably bounding values based on expert elicitation

- Leave margin

- Use natural analogues
  - Analogue research can be part of performance confirmation program

- Add/modify engineered feature to reduce importance of the FEP
  - E.g., drip shields added to mitigate groundwater flow uncertainty/heterogeneity issue

Important to identify these FEPs early.

  15   EPRI

Step-wise licensing
Post-closure reactor equivalents

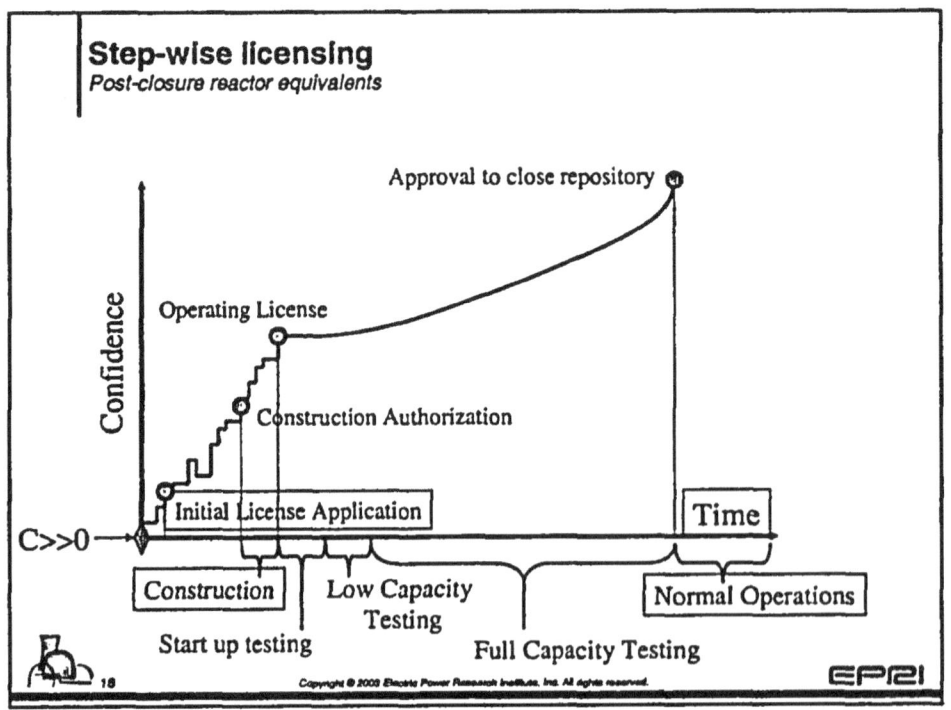

EPRI

Analogy to Reactor Licensing

PC similar to "Tech Spec" surveillance program for reactors

- Verify reactor equipment is operable

- "Limiting Conditions of Operation": what equipment must be operable and, if not, actions to be taken
  - In repositories, there are likely to be differing degrees of "inoperability"
  - Could be decades before "operability" needs to be restored or alternative action taken

17

Conclusion (1 of 2) : 'Big 3' EPRI PC Panel Long-Term R&D Issues

- Describe how a long-term R&D program (of which "performance confirmation" is only a part) provides enhanced 'confidence' in the future

- Considerations for activities to fit between each stage of repository development
  - SR, Construction LA, Construction authorization, Loading authorization, Closure LA
    - Widely different amounts of time between each
    - Commitments increase for specific FEPs

- Options for treating 'important' FEPs for which little additional information can be obtained over 25-300 years

18

Conclusion (2 of 2): Other Important Details

- Is appropriate 'baseline' information being collected at the right times?
- Establishing meaningful tolerance bands
- Identifying a clear (enough) end to the activity
- Prioritization in case of limited funding (or time)
  - Need to establish broadly based input on the criteria here?

 19

EPRI

6. GENERIC RESEARCH PERSPECTIVE ON LONG-TERM TESTING FOR PERFORMANCE CONFIRMATION - DEVELOPMENT OF AN INTEGRATED GROUND-WATER MONITORING STRATEGY

THOMAS NICHOLSON
NRC'S OFFICE OF NUCLEAR REGULATORY RESEARCH

Dr. Thomas Nicholson (NRC Office of Nuclear Regulatory Research) reviewed the ongoing development of an integrated groundwater monitoring strategy from a generic research perspective. The objectives of this research are (1) to develop technical bases for NRC staff evaluation of groundwater monitoring programs, (2) to couple monitoring to site characterization and facility performance assessment, and (3) to assess monitoring strategies to identify and support relevant alternative conceptual models of flow and transport. Other research objectives included (4) identification of hydrologic performance indicators, (5) development of a design strategy to collect monitoring data for parameter estimation, model calibration, and uncertainty analyses, and (6) accomplishing technology transfer to the NMSS staff.

Left blank intentionally.

Generic Research on an Integrated Ground-Water Monitoring Strategy

Thomas J. Nicholson 301-415-6268 TJN@NRC.GOV
Jacob Philip 301-415-6211 JXP@NRC.GOV
Office of Nuclear Regulatory Research
U.S. Nuclear Regulatory Commission

144th Meeting of the
Advisory Committee on Nuclear Waste
Rockville, Maryland
July 30, 2003

Outline

- Generic[1] Ground-Water Monitoring Needs
- Research Objectives
- Research Tasks
- Generic Applications
- Summary

[1] LLW, Assured Isolation Facilities and Decommissioning

Generic Ground-Water Monitoring Needs

- What, when, where and how to monitor for water flow and transport of contaminants

- Design monitoring systems to detect both current conditions and changes in system behavior that affect contaminant transport

- Develop database for identifying and quantifying causative mechanisms (e.g., events and processes)

3

Generic Ground-Water Monitoring Needs (continued)

- Identify potential for preferential transport pathways (e.g., features)

- Assess effectiveness of contaminant isolation systems (e.g., performance/degradation of engineered barriers)

- Data management, analysis, visualization and communication of monitoring data

4

Research Objectives

- Develop technical bases for NRC staff evaluation of ground-water monitoring programs

- Couple monitoring to site characterization and facility performance assessment (PA)

- Assess monitoring strategies for identifying and supporting relevant alternative conceptual flow and transport models

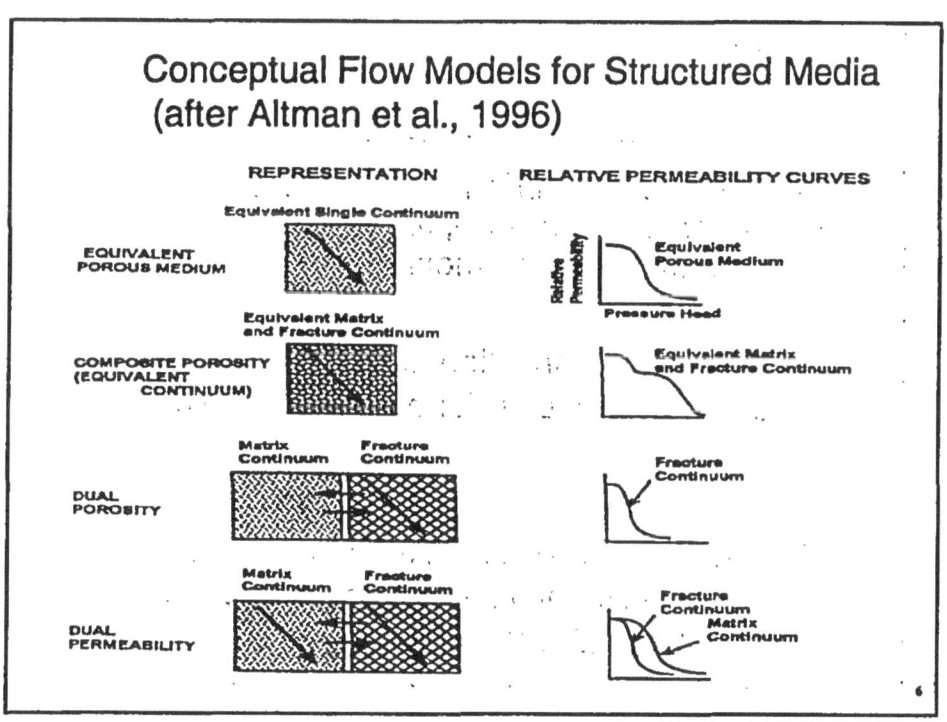

Conceptual Flow Models for Structured Media (after Altman et al., 1996)

Alternative Conceptual Models for Transport from Hanford Tanks

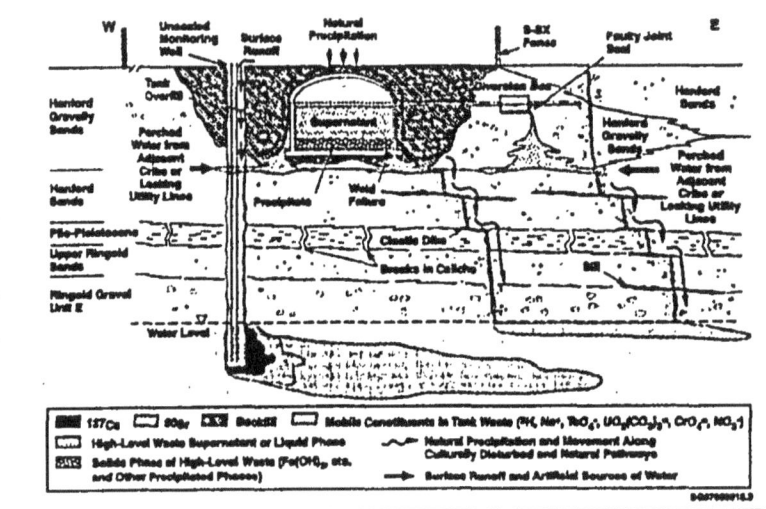

from Ward et al. (1997) after Caggiano et al. (1996)

7

Research Objectives (continued)

- Identify relevant performance indicators (e.g., water content, pressure, flux, contaminant concentrations) to be monitored

- Demonstrate connection between performance indicators and site performance as predicted by PA models

- Design strategy to collect monitoring data for parameter estimation, model calibration and uncertainty analyses

Research Objectives (continued)

- Update PA models using system monitoring data and analyses to generate new realizations of system performance

- Technology transfer to NMSS staff

Research Tasks

- Review and harmonize ground-water monitoring strategies presently used to evaluate nuclear & hazardous waste facilities

- Develop *Integrated Monitoring Strategy*

- Develop test plan for evaluating the *Integrated Monitoring Strategy* for a range of hydrogeologic features, events and processes

Research Tasks (continued)

- Test *Integrated Monitoring Strategy* by application to specially-selected monitoring datasets

- Technology transfer to NMSS staff

- Document and publish *Integrated Monitoring Strategy* and tested applications

Generic Applications

Provide practical information for:

- Understanding monitoring needs at sites to update and verify PA

- Identifying and evaluating alternative conceptual models related to causative mechanism (e.g., episodic recharge events) and its effects on transport

Generic Applications (continued)

- Estimating parameter and boundary conditions, and assessing uncertainty in PA models

- Coordination with participants in the MOU on multimedia environmental modeling research (*http: www.ISCMEM.Org*)

13

Summary

- Couple monitoring to site characterization and facility performance assessment
- Monitoring strategy to provide evidence for comparing and supporting alternative site conceptual models
- Ongoing NRC-funded research study is evaluating existing monitoring technologies
- Technology transfer to NMSS staff

14

Left blank intentionally.

7. WORKING GROUP ROUNDTABLE PANEL DISCUSSION ON PERFORMANCE CONFIRMATION

7.1 Summary of Panelist Comments

The six expert panelists (Whipple, Weart, Frishman, Bernero, Kessler, and Parizek) participated in a panel discussion that was moderated by Member Ryan.

Chris Whipple noted that 10 CFR Part 63 requires performance confirmation for all barriers that are classified as important to safety and that the PC work must be practicable. He considers there is potential conflict between the two requirements, and he thinks there is a possibility that DOE has not prioritized well and has failed to see the downside to classifying so many things as important to safety. Mr. McCartin (NRC) responded that DOE has some flexibility in deciding which barriers it will rely on. There is no numerical value given to describe the significance of barriers, but NRC would expect the DOE to look at the most significant barriers and apply most of the technical basis in DOE's safety case. In looking at PC, DOE would also be looking at the barriers it is relying on most. Dr. Whipple wondered what NRC would do if DOE identified a larger number of barriers than a reasonable person might technically believe are important. Would NRC rescue DOE from its own folly? Mr. McCartin replied that NRC is not there to "rescue" DOE. He referred to NRC's review plan for post-closure performance and noted that it emphasizes up front the identification of barriers important to performance. The intent is to tailor the NRC review to closely examine the barriers that DOE relies on the most. Generally, an NRC review focuses on what hasn't been considered or has been left out.

Robert Bernero (NRC, retired) observed that this is a classic problem in nuclear licensing involving the NRC. The applicants for a license are chronically looking for a prescriptive formula, "Tell me what I need to do so I can do it and you'll therefore give me a license." And the staff is chronically trying to give a description, an approach, but the responsibility for the logic and the supporting programs is the applicant's. That's a common problem, and especially so for DOE because the DOE is not accustomed to conducting its affairs as a regulated licensee.

B. John Garrick (ACNW Chairman) stated that the issue of classifying something as safety- or nonsafety-related is reminiscent of an analog used in probabilistic risk assessments, i.e., the "rocks in the pond" example. You have a pond that has a lot of rocks sticking out, and when you remove the biggest rock, the pond level goes down a level and some more rocks surface, and finally you remove enough rocks and the remaining rocks are small enough now that the surface doesn't significantly change as they are removed. That's what the performance assessment is supposed to give you. The answer to the question of whether or not it's important to safety is whether it makes any difference to the bottom line. If the performance assessment was competently prepared, there will be a road map that says "I'm not going to measure or worry about this particular rock because no matter what I do with it doesn't change the performance, it doesn't change the lake level of the "pond." If we have any confidence in our analysis at all, we have an inherent mechanism for classifying whether it's safety important or not, whether we need a particular barrier or not, whether it contributes to performance or not.

Steve Frishman (representing the State of Nevada) discussed, as an example, the parameter of matrix diffusion. Years ago the DOE had decided not to take credit for it because it was worth a relatively small percent of performance. It is also relatively unimportant in NRC's model. DOE seems to be reconsidering the potential contributions of such parameters. Mr. Frishman supported the idea that if a parameter is not worth a lot to performance for an applicant, to avoid an onerous review process, don't take credit for the parameter in the first place.

Richard Parizek (Penn State and Nuclear Waste Technical Review Board) stated that he was speaking as a private citizen rather than as a board member. He mentioned that some very valuable lessons were learned at the WIPP, that there is a real program there. There is an opportunity to understand how that program worked and why those decisions were made to include or not include certain testing efforts. There's a lot to be said about what we need to know about a site and about the characteristics of the site. For instance, what is assumed about climate in the TSPA model? Look at the Death Valley area (California) and the Mojave River drainage basin and we see in 10,000 years four major lake level stands in the basins. There were several periods of alluvial fan development, which really requires big triggering mechanisms to flush sediments down to generate the fans. So there's something about this weather story and about monitoring that might then say, "I'd better start looking underground because maybe this is a time when fast paths will kick in and this may have something to do with repository behavior." Dr. Parizek noted that from a science understanding point of view and confidence building point of view, some people wouldn't care where the money came from as long as performance confirmation got done. He discussed a number of possible monitoring activities, such as the placement of a monitoring well to monitor water chemistry and groundwater elevations and the drilling of magnetic anomalies to try to detect buried basalt flows.

Mr. Bernero asked, "What shall the program pursue in performance confirmation testing?" He noted that barriers should be tested, but unimportant barriers may not be. They may be set aside, but important performance assessment models may call for resurrecting. The key thing is to test models and the performance assessment. The performance confirmation program, the entire safety analysis, has to be a living system, a living document, learning and incorporating that learning and changing accordingly. It is important in any program to look at those things that provide important support for performance assessments, but that's not quite all you want to do. What is needed is to go beyond trying to measure things that can confirm that performance, and look broadly enough to find any holes or differences in models or assumptions that may surround those models and techniques that you believe to be correct. Usually surprises come in findings things that we didn't expect, and performance confirmation as a tool ought to be broad enough to look for those kinds of things.

Wendell Weart (Sandia National Labs, Senior Fellow) spoke about his WIPP experiences, and noted that DOE sometimes promised to do things that they weren't able to do. He expressed hope that performance confirmation wouldn't become a "shopping basket," that confirmatory activities would be selected carefully based on what is really important. Dr. Weart noted that this is a program that's going to be long enough that early on there may be intense interest and funding for it, but in future the funding may lag, making it a continuous struggle to implement the program. Regarding the use of conservative bounding arguments, Dr. Weart found from his WIPP experience that programs of long duration can be hurt by the fact that bounding conservatisms have been locked in, and are very hard to change after the fact. He advised not adopting these conservatisms unless it really is necessary.

Dr. Whipple also commented on the idea of avoiding bounding analyses and trying to be as realistic as one can be. Regulators find enormous comfort in being handed a bounding analysis with a lot of margin. There's little chance of that coming around and biting them. Dr. Whipple thinks this could similarly be true for a 9-million-page license application to the NRC. He noted that one aspect of a fully realistic analysis is it represents best understanding, best estimates with a kind of a 50-50 chance of being wrong in the non-conservative direction. This may be unacceptable in a politically charged, politically visible licensing process. As desirable as it would be to have a fully risk-informed approach through the licensing process, it would be a very risky strategy for an applicant to take. There is intellectual merit in a risk-informed approach, but the political reality of a licensing approach is the burden is on the applicant to prove that everything they say is either true or wrong in the safe direction. That is not fully compatible with being realistic and risk-informed.

Mr. Bernero responded that NRC, in its approach to a probabilistic risk analysis for reactor plants, made a concerted effort to be realistic, but approached realism from the conservative side of the field. There was simplification. For example, if conditions for adequate core cooling are lost, it was assumed that the core melted right away rather than try to mechanistically model the whole process. There was a very important reason why that could be done in a regulatory environment. Mr. Bernero noted that NRC consciously avoided regulating with a safety goal. It described a safety goal, one-tenth of 1 percent increment of background risk, etc., but did not regulate to the safety goal. It was intended for retrospective using performance assessments, or probabilistic risk assessments (PRAs), that were as realistic as they could be made. The big difference regarding high-level waste is that the regulation is fundamentally based on the performance assessment. It's not a safety goal, it's a condition of acceptability. The real question is trying to understand the margin, trying to understand what confidence you can have in those results, and trying to understand barriers that right now may not be very important, but if the principal barrier of the package, etc., fails, they become very important. Mr. Bernero considers there's a fundamental difference in NRC history in that regard.

Mr. Frishman responded that some people have suggested that performance assessment should be an exposure of what you know. It should be possible to accurately characterize and quantify what you don't know. On the other hand, a performance assessment has to be used for compliance because that's what the rule says. Mr. Frishman suggested there may be the need to develop an expectation for two kinds of performance assessments. One of them will meet the need required by the rule to demonstrate what you know, and the other will show compliance based on an assessment of a demonstration of what is known.

Dr. Weart commented that one can't always judge in which direction conservatism exists. And if you're smart enough to have thought of everything in advance and say, "I'm never going to have any surprises," then perhaps you're okay. But Mr. Weart advised that if you don't have to rely on bounding analyses, don't, but there are times when perhaps it's all right. But it can come back to haunt you.

Mr. Bernero commented on DOE's decision analysis for selecting the PC portfolio. He found the decision analysis process difficult to track but clear, and thought it was very well done, a logical process, clearly tracked, and producing a reasonable result. However, he found some of the characterization of portfolios A through K to be unclear. Portfolio A was identified as the minimum needed to satisfy the regulator. Mr. Bernero felt that wouldn't be right because that would be the minimum necessary. The applicant would be saying, "I know all I have to do is tell them this, and that's enough to satisfy them." He interpreted DOE's selected portfolio C "plus" as representing the best judgment of the applicant. It is DOE's responsibility to come up with the right performance confirmation, to show how they're going to satisfy the regulatory requirement. NRC would review that, and that sounds like the right way to choose a portfolio. Mr. Bernero commented that the NRC avoids, and should avoid, overly prescriptive regulation. NRC shouldn't give DOE a prescriptive description of what the performance assessment should be. But NRC should develop alternative models of their own. They should be giving descriptive analyses that say what the performance confirmation ought to be.

Dr. Garrick stated that the regulator is never the expert on the system being licensed that the operator-owner is. Never. No matter how many regulations, no matter how many lawyers the regulator has, the regulators do not know the system as well as the owner, operator, designer, builder, or whoever. The perspective should be that the most expert group in the world on that system is completely satisfied that the system is safe. They shouldn't even think compliance—they should think totally from the standpoint that it's safe, and then let the licensing people worry about whether they've complied with the regulations. Mr. Bernero agreed that the regulators are not the ultimate experts, and regulations cannot be so prescriptive as to have specific solutions to problems. But they can require a competent quality assurance (QA) program. He remembered signing a letter on July 31, 1989, to the Yucca Mountain program that said, "This won't wash. Your site characterization plan — we have two objections to it. You don't have an adequate QA program, and you don't have an adequate design control process." NRC did not tell them what those processes had to be. But DOE was told that what they had didn't "cut the mustard." The regulator can't pose as the expert, but the regulator can say, "You don't meet the standards or evidence. You don't show evidence of sufficient safety or competence in an area."

Dr. Kessler commented that, since Yucca Mountain is a first of a kind project, it's probably okay for there to be a bit more guidance from NRC, given that this is the first one out of the starting block. This doesn't mean a lot more specification, but perhaps some clarification of the relative importance of supporting the barrier analysis versus just supporting the overall performance criteria. Perhaps DOE needs to back up and add a little bit more on the realistic side to provide some insight on how much margin they're providing in their compliance-based assessment.

A number of participants discussed the manner in which NRC would review DOE's performance confirmation plan, which would include discussions in public meetings about what is reasonable for the program to include. Dr. Kessler commented that this dialogue needs to begin now. Dr. Parizek commented that "it's not collusion, it's trying to be efficient with the use of everybody's time and getting to the end point. Mr. Frishman expressed the concern that it will be a very difficult situation if the applicant and the regulator are essentially negotiating the meaning of the regulation. He suggested that there is no real precedent for this. Mr. Frishman felt that to do the informal negotiation prior to licensing could be antithetical to an accountable regulatory system. Dr. Kessler responded that there seems to be plenty of precedent for the regulator and the applicant to have discussions on a generic basis. He gave examples of very quantitative, specific interim staff guidance that grew out of technical discussions in publicly noticed meetings where the applicants and the regulator sat down and talked about a technical detail. Dr. Kessler considers that this happens all the time, and it's done in public meetings with that kind of level of discussion.

7.2 Closing Remarks by Keynote Speaker Chris Whipple (Environ) During the Roundtable Panel Discussion

DR. WHIPPLE: As I listened to the last day and a half, what came across for me is an important point with respect to performance confirmation. Performance confirmation is to be done for things that are important to safety. We've clearly heard that 10 CFR 63.131 through 134 requires performance confirmation for all barriers that are classified as important to safety, as opposed to being safety significant in a performance assessment sense. And it has to be practicable.

I see the potential conflict between the first two requirements, and it may well be that DOE has simply extended the definition of barriers important to safety beyond the logical stopping point with the consequence that now you need to do performance confirmation on things like gravel in the bottom of the drift, which to most of us might not be seen as terribly important to safety. This is a consequence of semantics and a poor choice by DOE not recognizing a downside to classifying so many things as important to safety.

But I would like to hear, particularly from the NRC staff, if they think there is a substantive requirement for importance to safety somewhere else in 10 CFR Part 63 other than in the 131 to 134 link that might be a basis for not doing some things that appear to be pretty low valued. That to me is the central question that's emerged after a day of listening to this.

Left blank intentionally.

8. PUBLIC COMMENTS

Day 1

Ms. Judy Treichel (Nevada Nuclear Waste Task Force) noted that one of the things that would provide some public comment would be to know that we could get the presentations with not just the odd-numbered pages, because I like to write on them and I don't like getting them later, and I still want to get one of Debbie Barr's last handouts, because that was never out there. So that's just a little QA problem that pops up from time to time.

I think the whole discussion has been really strange. I was part of or attended and made a comment at the December meeting that was mentioned here about performance confirmation, and the fact that as we've been hearing all through these presentations, that there has to be a performance confirmation that started during site characterization. Obviously if the Department is now in the process of coming up with one, it wasn't there during site characterization.

If we're working on Rev. 2 of the performance confirmation plan, there had to be a Rev. 0 and a Rev. 1, and I never got those, and I was supposed to be getting them, and I suppose there will be something on there that happened already so they could say that they had something, but this performance confirmation really looks like something in its infancy.

It lends itself to comments like Chris Whipple made when he said that the word, "confirmation" could indicate an overconfidence or could send the wrong message. Well, what we were told as the public, the people who are supposed to be getting all of this new confidence, was that if there was too much uncertainty, if you weren't really confident, if the site really wasn't shown to be doing what it had to do, it wouldn't happen. So I'm not sure that a performance confirmation program's going to give us what should have already been there. I doubt that it would. But we seem to be in the very first steps of something.

And then once you get to this point where you're just putting it together, we're really nervous about things that have to happen in the future, like the $8 billion worth of titanium that has to get thrown in there as drip shields—it's promised now but has to be paid for later. A lot of this program is going to have to be paid for later. So is there going to be some sort of a financial bond that goes with this, some kind of a promise where you've got the money in the bank and you know that it's going to happen? Because it doesn't always happen.

And as Debbie Barr said, some activities could be deleted or replaced. Well, I'm sure they could. When we came up with the KTIs, each one of those at the time that it was put down as an action item or as an issue, it had to be resolved and it was important. And now we're seeing some of them becoming less important or being able to be shuffled off or something. This does appear to be a collection of things that would be much handier to be able to do later if there's money, if there's time. And if it had already been done during site characterization, which I and a lot of Nevadans believe it should have been done, then we wouldn't be worried about whether or not there would be money to do it.

And I'd also like to know if there's any possibility that things could stop if in fact this laundry list of new scientific marvels like the remotely operated vehicles and so forth don't come through or if when they do it's a problem to get them to work with all that heat or radioactivity. Is any of this stuff going to be shown working? The word "retrievability" is always thrown around, and I don't think that that would ever be demonstrated in any way that it should be. But even these

191

things that are now going to be part of a program that's required really need to be proven that they can happen and that they will be paid for.

Day 2

Ms. Judy Treichel (Nevada Nuclear Waste Task Force) was very concerned that the PC program is not far better defined at this time. She said this is one reason that the site recommendation and sufficiency letter were premature. She considered Yucca Mountain to be a project forced on an unwilling host. These are people (Nevadans) who do not like the idea of being the host for the repository, and they really don't like DOE. These nuclear-testing people killed us once; we're silly if we let them do it again. And Nevadans have been told for years and years, you don't have to like DOE, you don't have to trust DOE, because you've got NRC. And NRC is going to come in here—they will only license this thing if it's absolutely safe, and NRC will take charge of your safety, your health, and your well-being. So be clear about that. That's what has been told to Nevadans, and that's what their expectations are. And you've got people who are very nervous. We don't want to see compromises. You already know the lay of the land in Nevada. But don't let this thing become some sort of an excuse. She is eager to see what performance confirmation winds up being. But she doesn't want it to be something that just hangs over everybody's head.

Dr. Elzeftawy (consultant to Las Vegas Paiutes) said that the performance confirmation program needs to be simple but beautiful for the people to have confidence that this program is on track and is applicable. He noted that we, as scientists, discuss these issues but the public has some common sense and needs to understand the simplicity of performance confirmation. The NRC has the responsibility of looking at it. But NRC needs to come to a focal point, and the focal point is to make it simple and understandable to most people.

Dr. Elzeftawy said that, as a person who has left the program and then saw a couple of things during the last year or year and a half, he is reminded of a saying, "The more things change, the more they stay the same." It seems that we are back into the discussions of 1982, 1983, 1984, when he joined the NRC. We are still more or less standing still. How much progress has been made? DOE may spend about $2 or $3 billion, which we spend now in less than 3 weeks. What do we have to show for it? You need to look at that point and make it public, because this is a public program. You also need to hold more meetings in Las Vegas. He doesn't think anybody in Las Vegas or in the State of Nevada will come up with $3,000 in his pocket to come here to attend your meeting and stand here and give you the public opinion. Hold many, many, meetings, as many as you can, not in the NRC building, and not over there. Come to the public in Nevada. Dr. Elzeftawy noted that Yucca Mountain is a very important program to the Nation, and that there's a lot of responsibility placed on the Department of Energy (DOE) and the NRC. He said that there will be a lot of political heat on the Commission, but some day they'll have to vote. So there's going to be a very tough political situation—and a hard decision to make. But NRC lays down the ground rules and the information that will be used by the people and the Congress and others.

C FORM 335
)04)
MD 3.7

U.S. NUCLEAR REGULATORY COMMISSION

1. REPORT NUMBER
(Assigned by NRC, Add Vol., Supp., Rev., and Addendum Numbers, If any.)

BIBLIOGRAPHIC DATA SHEET

(See Instructions on the reverse)

NUREG/CP-0190

ITLE AND SUBTITLE

'roceedings of the Advisory Committee on Nuclear Waste

Vorking Group on Performance Confirmation Plans for the Proposed Yucca Mountain ligh-Level Waste Repository

3. DATE REPORT PUBLISHED

MONTH	YEAR
December	2004

4. FIN OR GRANT NUMBER

UTHOR(S)

leil M. Coleman

6. TYPE OF REPORT

Technical Report

7. PERIOD COVERED *(Inclusive Dates)*

ERFORMING ORGANIZATION - NAME AND ADDRESS *(If NRC, provide Division, Office or Region, U.S. Nuclear Regulatory Commission, and mailing address; If contractor, rovide name and mailing address.)*

\dvisory Committee on Nuclear Waste

J. S. Nuclear Regulatory Commission

Vashington, DC 20555-0001

PONSORING ORGANIZATION - NAME AND ADDRESS *(If NRC, type "Same as above"; If contractor, provide NRC Division, Office or Region, U.S. Nuclear Regulatory Commission, ind mailing address.)*

iame as above

SUPPLEMENTARY NOTES

ABSTRACT *(200 words or less)*

This report contains information presented at the Advisory Committee on Nuclear Waste (ACNW) Working Group Meeting on Performance Confirmation Plans for the Proposed Yucca Mountain High-Level Waste Repository held on July 29-30, 2003. This report summarizes the presentations given to the Committee, along with the presentation materials and selected discussions among the participants.

The purposes of the working group meeting were (1) to increase ACNW's technical knowledge of plans to develop and conduct performance confirmation work for the proposed Yucca Mountain repository, (2) to understand NRC staff expectations for performance confirmation, (3) to review examples of performance confirmation work being planned, (4) to identify aspects of performance confirmation that may warrant further study, and (5) to complement a previous working group meeting on performance assessment.

KEY WORDS/DESCRIPTORS *(List words or phrases that will assist researchers in locating the report.)*

Yucca Mountain
10 CFR Part 63, Subpart F
decision analysis
groundwater monitoring
performance confirmation plan
performance confirmation program
U.S. Department of Energy
geologic repository
high-level radioactive waste
subsurface conditions
natural and engineered barriers

13. AVAILABILITY STATEMENT

unlimited

14. SECURITY CLASSIFICATION

(This Page)

unclassified

(This Report)

unclassified

15. NUMBER OF PAGES

16. PRICE

Printed
on recycled
paper

Federal Recycling Program

www.ingramcontent.com/pod-product-compliance
Lightning Source LLC
Chambersburg PA
CBHW080241180526
45167CB00006B/2374